Günther Dichatschek

Education in Europe 5

Günther Dichatschek

Education in Europe 5

Aspects of ecological education

ScienciaScripts

Imprint

Cover image: www.ingimage.com

This book is a translation from the original published under ISBN 978-3-639-63332-0.

Publisher:
Sciencia Scripts
is a trademark of
Dodo Books Indian Ocean Ltd. and OmniScriptum S.R.L publishing group

120 High Road, East Finchley, London, N2 9ED, United Kingdom
Str. Armeneasca 28/1, office 1, Chisinau MD-2012, Republic of Moldova, Europe
Managing Directors: Ieva Konstantinova, Victoria Ursu
info@omniscriptum.com

Printed at: see last page
ISBN: 978-620-8-51644-4

Education in Europe 5

Aspects of ecological education

Günther Dichatschek

Table of Contents

Preliminary remark

To develop a basic knowledge of ecology requires an interdisciplinary approach in civic education, because the subject matter arises from a wide variety of subject areas. For the author, the broad possibilities of further education in the tertiary and quaternary education sector have arisen in the course of time.

Personal points of reference are therefore dealt with in the following chapters of the study.

Ecology is understood as knowledge of the relationships of individuals or institutions to the surrounding environment, as in the following.

- Climate change and climate protection,
- Alps and Alpine ethics,
- Sustainability,
- Mobility and sustainability,
- Learning cultures in continuing education and
- exemplary economic and environmental ethics.

The author is also grateful for the input from colleagues.

Observing the technical literature is an essential part of basic knowledge.

Basic knowledge is understood as the basis for future areas of knowledge arising from the subject matter.

I Climate change and climate protection

Introduction

The increasing interest in climate change and the resulting consequences is important for contemporary civic education because it has consequences for

- nature and society,
- Options for action,
- scientific climate research and
- consequences for political action (see DICHATSCHEK2017a).

Climate change is here, changes in the form of extreme weather conditions around the globe are perceptible. The rise in temperature and humidity is being felt by humans, animals and plants.

- These and other phenomena will intensify and change living conditions.
- They influence the future and raise fundamental questions that civic education must address. Attention must be paid to the subject-relatedness of the subject at school or the subject combination with geography and economics, biology and environmental studies, history - social studies - political education, religion, ethics and health education as well as the topicality in adult education and further education (see DICHATSCHEK, 2017c).
    - Can we adapt to the changes?
    - How can the observations be categorised?
    - What causes can be attributed to the changes?
    - Are forecasts possible?
    - Are there opportunities for personal action?

The starting point for the following considerations are

- the school curricula for the subjects in question,
- the curriculum in the teacher training programme for history - social studies - political education and

- the need to organise contemporary adult education.

For the author, the necessity of dealing with the topic arises from the

- Completion of the 10th university course in political education at the University of Salzburg-Klagenfurt,
- Completion of the 7th University Programme Intercultural Competence/ University of Salzburg,
- Teaching activity in didactics of political education/teaching programme history-social studies-political education/University of Salzburg,
- Teaching activity in adult education,
- Completion of the online course "Climate M00C7 October-November 2017 > http://mooin.oncampus.de > Climate MOOC and
- Discussion of the specialised literature.

1 Predictability of climate change

As a complex entity, the climate cannot simply be predicted. Science requires computer models in particular as a tool for assuming climate developments and forecasts. The assumptions are summarised as scenarios. They contain possible developments in the coming decades (cf. population figures, technological progress, limitation of greenhouse gases) (cf. MADRY - FISCHER 2014, 1-2/4; PLÖGER - BÖTTCHER 2016).

- Man-made climate change has been proven by measurements since the 1950s. It is essential to assume three scenarios with physical, scientific and human data.
- Greenhouse gases play a central role in this. The gases contain substances in the atmosphere that cause the greenhouse effect on Earth.
    - Short-wave radiation from the sun is converted into long-wave radiation at the earth's surface and radiated back into space.
    - Greenhouse gases act as a barrier in the Earth's atmosphere, reflecting back radiation and therefore helping to store energy and heat on Earth.
    - The greenhouse effect is essential for life on earth. Without this influence, the average air temperature would be -18 degrees Celsius. The natural greenhouse effect leads to warming to around 15 degrees Celsius.
    - The effect is now so pronounced that we can speak of man-made global warming due to the increased concentration of greenhouse gases. What is special is the speed at which the warming has been taking place since industrialisation in the mid-19th century.

2 Consequences of climate change

Austria is (also) affected by climate change (see MADRY - FISCHER 2014, 2/4; GREENPEACE 2019).

- The alpine regions (risk of severe weather - landslides,

flooding from mountain streams; melting ice) and water shortages in the arid climate zone must be taken into account.

- In the last 100 years, the average annual temperature has risen by around 1 degree Celsius, and an increase of 3 degrees Celsius is expected in the next 100 years.
- Precipitation has increased in spring and winter, while less rain falls in summer and autumn. The impact on the environment, animals, plants and humans is considerable.
- A distinction must be made between microclimatic, regional, general meteorological and macroclimatic phenomena.

The rise in temperature in southern regions of the world is leading to water shortages, severe storms (tornadoes, hurricanes, typhoons, cyclones, massive monsoon phenomena) and the melting of ice at the poles (see PLÖGER - BÖTTCHER 2016, 153-157). A change in sea temperatures can be observed.

The Intergovernmental Panel on Climate Change (IPPC) expects an increase in the number and intensity of heat events and extreme storms in our climate zone.

- By the end of the century, the number of summer days with > 25 degrees Celsius will double and the number of hot days with > 30 degrees Celsius will triple.
- Regionally, the trend may intensify.
    - Problems arise here in urban centres (water supply, health hazards due to heat).
    - In rural areas, crop failures are increasing (storms, water supply - heat waves - famines and malnutrition).
- In tourism, a reduction in the number of snow days is significant in the low mountain regions. Warmer and longer summers, especially in coastal regions, offer the chance of more guests and therefore better sources of income.
- In nature, attention must be paid to changes in species diversity. Immigration of species from warmer regions and migration to higher regions are signs of change.

For the Central European region, it should be noted that

- the costs of climate change will be an investment in the future (see PLÖGER - BÖTTCHER 2016, 62-68),
- the energy transition is a necessity (see PLÖGER - BÖTTCHER2016, 74-83),
- our big cities will be a hotbed of heat (see PLÖGER - BÖTTCHER 2016, 115-118) and
- climate change is accelerating the extinction of species (cf. PLÖGER - BÖTTCHER2016, 129-133).

IT - Note

Biodiversity Strategy 2030 Austria > https://orf.at/#/stories/3297397/ (12.12.2022)

3 Fields of action for climate change

The following fields of action are affected by regional climate change (cf. MADRY - FISCHER 2014, 3/4).

- Buildings - exposure to extreme weather, requirements for building cooling
- Water management - allocation of capacity limits, water pollution
- Health - spread of new infectious diseases and vectors, heat stress, water quality pollution, malnutrition due to hunger (cf. WEINGÄRTNER 2014)
- Financial management - increase in costs for insurance and health insurance, risks for investments (cf. water protection, protective structures)
- Transport - Transport infrastructure during weather events
- Energy - increase in demand (cf. cooling units, cooling water, loss of energy)
- Regional and spatial planning - conflicts of land use, new requirements in spatial planning

4 Climate policy

Political education includes conflicts that have arisen in the subject matter and ways of overcoming them.

The starting point for the considerations is the state of knowledge of the MCC (Mercator Research Institute on Global Commons and Climate Change/ Berlin) and Potsdam Institute for Climate Research (PIK) in conjunction with the Intergovernmental Panel on Climate Change (IPCC) (cf. EDENHOFER-JAKOB2017, esp. 109-120).

Climate policy deals with

- understanding of the climate problem and climate change,
- the possibilities of averting and managing risk.
- It is about taking stock of the situation and the goals and paths of climate policy.
- international agreements,
- the climate policy of the European Union and
- urban development and climate protection.
- The role of science as policy advice is essential.

IT - Notes

- https://orf.at/stories/3296904/ (8.12.2022)
- https://archive.newsletter2go.com/?n2g=cc35u22b-g39z6tuq-k13

(10.12.2022)

4.1 Climate problem - Climate change

Greenhouse gases are stored in the oceans, the atmosphere, soils and forests ("global sinks").

- The stock of greenhouse gases is increasing due to the burning of fossil fuels, deforestation, land use and industrial processes.
- The concentration of greenhouse gases changes the Earth's radiation budget.
- The reflected sunlight increasingly remains in the atmosphere

in the form of heat, which is why the global mean temperature increases. Local climatic conditions are also affected, as is the circulation of air and water.

An increase in the global mean temperature poses risks to living conditions on Earth. The Intergovernmental Panel on Climate Change has categorised the risks into classes (see EDENHOFER - JAKOB 2017, 10-13).

- Threats already fall into the first class at + 1.5 degrees, for example for coral reefs and ecosystems in the Arctic.

- Heat waves and tropical storms fall into the second risk category.

- Regional risks and loss of income are summarised in the third risk class.

- The fourth risk class comprises economic damage to the global economy.

An increase in temperature of just + 0.8 degrees leads to changes in precipitation (lower agricultural yields), acidification of the oceans (lower fishing yields) and a reduction in the quality and quantity of available water resources.

The climate problems outlined above point to massive challenges for health policy in the 21st century, such as cardiovascular diseases, polluted water and water shortages, food shortages and the spread of epidemics.

Adaptation strategies relate to irrigation systems, higher dykes, coastal protection and resilient infrastructure, whereby it must be borne in mind that unchecked climate change will lead to regional cost increases and ultimately to their limits.

Waves of migration can occur due to a lack of livelihoods (cf. crop failures, floods and hurricanes).

Making a decision is a complex process. Assessing the future is also an ethical process, as it involves clarifying the equal treatment of all generations.

Further reading

Rahnstorf S. - Schellhuber H.J. (2007): Climate change. Diagnosis, prognosis, therapy, Munich

4.2 Inventory

Understanding the challenges involves taking stock of climate policy (see EDENHOFER - JAKOB 2017, 18-37).

4.2.1 Development of emissions

Based on data from the "Climate Analysis Indicators Tool" (CAIT), the CO2 equivalents are used to take all greenhouse gases - including methane and nitrous oxide - into account.

- The so-called "global warming potential" is used as an index to compare the different greenhouse gases.
- For each greenhouse gas, the climate impact for a certain period of time is expressed in comparison to CO2. Methane has about 28 times the climate impact of CO2, but remains in the atmosphere for less time. Although CO2 has a lower warming potential, it remains in the atmosphere for a long time.

Annual greenhouse gas emissions rose by 44 per cent between 1990 and 2014.

- The reasons for this are economic growth in developing and newly industrialising countries, the increasing use of coal in the electricity sector, the burning of fossil fuels and the emission of CO2 in industrial processes (such as the chemical industry).
- In recent years, developing countries have overtaken industrialised countries. China is now the world's most significant emitter, with Brazil and Indonesia leading the way in terms of land use as a source of emissions with rainforest utilisation.

Globalisation has also led to emissions being exported.

- The metric used is "consumption-based emissions". They measure the emissions generated in the global upstream supply chain of the goods and services consumed.
- They are calculated by adding the emissions contained in imports to the production-based emissions.

- Emissions cannot be measured and are therefore estimated via the production processes.
- The importance of this process is demonstrated by the fact that internationally traded products account for a quarter of global emissions.
- The reasons for this are different energy generation technologies, unbalanced trade balances and specialisation in emission-intensive products.

It is clear that the distribution of the burden of climate protection in the form of greenhouse gas emissions cannot be decided solely on the basis of consumption- or production-based emissions; political and ethical issues also play a key role.

4.2.2 Population and economic growth

In the last 200 years, population growth has increased more than sevenfold, which in turn has increased per capita income more than tenfold (see EDENHOFER - JAKOB 2017, 27).

The UN predicts that the world's population will reach another two billion people by the end of the century.

- The global economy will also grow strongly in the coming decades.
- Population and economic growth is reflected in the increase in emissions.
- Although energy utilisation has become more efficient as a result of the structural change from industry to services, this development is not enough.
    - It can be assumed that fossil fuels will remain dominant despite the use of renewable energies and efficiency technologies.
    - This can be seen in the development of developing and emerging countries, which are imitating the development of industrialised countries and are therefore unable to decouple energy consumption from economic and population growth.
- Such a decoupling can only be observed with a high per capita

income (cf. energy consumption in Germany in 2013 around 9 per cent lower than in 1990; UK 5 per cent lower; but France increase of 13 per cent).

The conclusion is that there are

- requires a suitable political framework,
- reduction targets are to be accepted,
- A sufficient energy supply is a basic prerequisite for economic development,
- energy is also provided by hydropower, wind, solar energy, biomass and nuclear energy,
- low-emission energy sources are often more expensive (cf. the argument in developing countries that education and social systems cannot be financed in this way and that the polluters [industrialised countries] must reduce their emissions first).

4.2.3 Supply of fossil fuels

With the oil crisis in the 1970s, the industrialised countries gained their first experience of the "limits to growth". In 1972, the Club of Rome predicted the imminent end of fossil fuels. However, it turned out to be an illusion that renewable energies would subsequently prevail. On the contrary, the rise in oil prices led to the search for new and undiscovered deposits, some of which were difficult to access (see EDENHOFER - JAKOB 2017, 32-33).

In addition, the "fracking process" led to a rapid increase in oil production in the USA and to a fall in the global market price from around USD 120 in 2009 to USD 35 at times in 2016.

- Of interest is the high capacity of the process in Canada, which would be comparable to the reserves in Saudi Arabia. Extraction would cause massive damage to the climate and the local environment.
- As a result, the supply of gas is also increasing. In the USA, for example, the electricity supply has increasingly been switched to natural gas. The lower price also results in competition with the price of coal.

- However, the lower demand for coal reduces its price, making more coal exports of interest. The effect of climate-friendly natural gas is therefore dwindling.
- There is a risk that up to 20 per cent more methane emissions will escape as a result of fracking. This means that the emission levels of natural gas are roughly comparable to those of coal.

4.2.4 Deforestation - Land use

Deforestation of rainforests and the resulting land use are the second largest source of emissions. The largest share of emissions comes from draining peatlands and clearing forests (see EDENHOFER - JAKOB 2017, 33-35).

- Peatlands store large quantities of CO2, which are released during melioration.
- Between 2000 and 2013, around 1.5 million square kilometres of forest were destroyed.
- Deforestation does not only occur in the southern hemisphere (on a massive scale in Brazil and Indonesia; cf. increased use as agricultural land, use for road construction, reduction in transport costs or wood use and colonisation), but also in the northern hemisphere in Russia and Canada. The consequences are climate change, soil erosion, increased pest infestation and forest fires.
- Compensation is provided by targeted reforestation (see Canada).

4.2.5 Renewable energies

With the rise in energy efficiency, renewable energies have been increasingly expanded (see EDENHOFER - JAKOB 2017, 35-37).

- Greater energy efficiency replaces labour and capital with energy (cf. low energy prices increase product demand and thus energy consumption).

- The costs for wind and photovoltaics are therefore of interest, and thus lower emissions. Depending on location and expansion rate, the costs remain even more expensive than fossil energy sources (cf. the subsidies for the pricing of fossil energy sources).
- Renewable energies already account for 13 per cent of global primary energy. Biomass such as firewood and dung make up half of this, while hydropower accounts for 22 per cent.
- Fossil fuels still account for almost half of new installations. Coal is cheap and plentiful.

Further reading

Sinn H.W. (2008): The green paradox. A plea for a climate policy free of illusions, Berlin

4.3 Goals and paths

In its most recent assessment report, the Intergovernmental Panel on Climate Change determined that the global mean temperature in 2100 depends on the cumulative carbon budget. The carbon budget adds up all emissions in the 21st century and determines the rise in the global mean temperature in an irreversible way (see EDENHOFER - JAKOB 2017, 37; hereinafter 37-62).

The political consequence is that if the 2-degree limit is to be adhered to, the atmosphere must be understood as a limited dumping ground for greenhouse gases. So far, this has not been the case. Utilisation rights have not been defined, so there is a risk of overuse. Only clear guidelines will make the atmosphere a global common property.

The 2015 Paris Agreement set the 2 degree limit as a target. It would be desirable to limit the increase to 1.5 degrees. Emissions should therefore only be emitted to the extent that oceans, land and technical systems remove them from the atmosphere.

In order to determine the optimum stabilisation level of the global mean temperature, it would be necessary to compare the costs of climate protection with the expected damage from climate change on the one hand and the costs of adaptation on the other. The cost-benefit analysis is only possible if all the consequences of climate change and the probability of their occurrence are known and a standardised assessment key is available.

In setting the 2 degree target, assumptions are made that cannot be derived solely from the biophysical consequences of climate change. The prerequisite is, for example, that

- climate impacts can be limited to an acceptable level, provided that this limit is adhered to,
- avoidance and adaptation is compatible across generations,
- economically and technically achievable without compromising sustainability goals.
- Ethical duty presupposes ability.

The following factors serve to reduce greenhouse emissions and are the subject of political endeavours and thus of political education. They are listed below in keywords and serve current considerations (see EDENHOFER - JAKOB 2017, 46-62).

- Use of technical innovations to reduce energy consumption and increase energy efficiency,
- Reducing electricity costs - increasing renewable energy - interconnecting the electricity grids,
- Reduction of emissions in the transport sector,,
- Decarbonisation in the industrial and agricultural sectors,
- Costs and risks of climate protection,
- Discussion about a growth waiver,
- Adaptation strategies in connection with
- Aspects of flight and migration movements.

Further reading

Stern N. (2006): The Economics of Climate Change. The Stern Review, Cambridge University Press

WBGU/ German Advisory Council on Global Change (2011): World in Transition. Social Contract for a Great Transformation http://www.wbgu.de/hg2011/ (16 June 2018)

4.4 International agreements

The agreements of Rio, Kyoto, Copenhagen and Paris are addressed below. They serve global agreements to prevent or minimise climate change (see EDENHOFER - JAKOB 2017, 75-81).

The starting point for international climate policy is the United Nations Framework Convention on Climate Change (UNCCCC) adopted in Rio de Janeiro in 1992.

- The aim of the convention is to prevent climate change.
- Further development in the context of globalisation takes place via annual summit meetings of the Conference of the Parties (COP).

The Kyoto Protocol of 1997, which came into force in 2005, obliges industrialised countries (Annex I countries) to reduce greenhouse gas emissions.

- According to this, emissions are to be reduced by 5.2 per cent below the 1990 level between 2008 and 2012.
- The Kyoto Protocol assumes that industrialised countries are the main cause of climate change and must reduce their emissions.
- In 2011, the countries that ratified the Kyoto Protocol were only responsible for 13 per cent of global greenhouse gas emissions (cf. the absence of the USA).
- To ensure cost-effective climate protection, emission allowances were allocated to the countries with emission reductions.
    - These were traded between the industrialised countries, with the price based on supply and demand.
    - As there was no price signal, emissions trading ceased.

The Kyoto Protocol did not succeed in reducing emissions globally, distributing emission rights fairly and achieving a price on the emissions market that would influence investments.

In 2009, a follow-up agreement was negotiated at the climate conference in Copenhagen. The "top-down approach" as a definition of emissions reduction and distribution among the members (burden sharing) was to be continued (cf. the rejection by the USA, China and many developing countries).

In 2015, the Paris Agreement overcame the division between industrialised, developing and emerging countries. Avoiding climate change is now considered a joint task. The burden-sharing has not been clarified in this agreement either. The agreement consists of three pillars,

- a formulation of the long-term goal - limiting the global mean temperature to 2 degrees,
- voluntary commitments - submission of national plans (Nationally Determined Contribution/NDC) without the possibility of sanctions and
- the agreement of several multilateral instruments - climate financing of at least USD 100 billion per year and the possibility of international emissions trading with an as yet unresolved design (as of 2018).

One transfer mechanism relates to forest protection (REDD+ mechanism/ Reducing Emissions from Deforestation and Forest Degradation). Compensation is provided if deforestation is reduced.

- Little attention is paid to the fact that landowners benefit from the transfer payments, as forest protection generally leads to an increase in land prices for agricultural land.
- Unequal distribution leads to an increase in land prices and thus to regional economic inequality with a tendency to impoverish small farmers.

Further reading

Hulme M. (2014): Climate change: Why there are no simple solutions to the greatest challenge, Munich

4.5 EU climate policy

The European Union (EU) wants to reduce its emissions by 40 per cent by 2030 compared to 1990, increase renewable energies by 27 per cent and increase energy efficiency by 27 per cent (see EDENHOFER - JAKOB 2017, 85-86).

- By 2050, the EU is aiming to reduce emissions by 80 per cent compared to 1990. This political target is not a legally binding commitment.
- To achieve these goals, the EU is influencing European emissions trading, functioning energy markets, instruments to increase energy security, the promotion of renewable energy and the integration of the European electricity market.
- Emissions trading as a central instrument comprises three components,
    - the determination of the upper limit,
    - the issuance of greenhouse gas emission allowances, and
    - the trading of these certificates, which should cover as many sectors as possible (cf. the lack of emission caps in the transport sector).

IT - Note

http://www.orf.at/stories/2443299/ > EU - States climate targets (18.6.2018)

4.6 Urban development and climate protection

The following aspects can combat major problems, such as

- the utilisation of the existing infrastructure and
- give priority to possible alternatives (e.g. public transport, kilometre-based charges for road use).
- In terms of transport policy, an efficient local transport system is favourable.
- Increased importance should be attached to spatial planning (e.g. a balance of building and green spaces).
- To avoid high inner-city property prices, a land tax to finance environmentally friendly infrastructure is advocated (cf. the financial inefficiency of a building tax) (cf. EDENHOFER - JAKOB2017, 101-102).

Further literature on 4.6

Plöger S. - Böttcher Fr. (2016): Climate facts, Federal Agency for Civic Education, publication series vol. 1734, Bonn, 115-118

4.7 Science and climate policy

Climate science has been able to help set the climate policy agenda in recent years. A fictitious look back from the future shows this (see ORESKES - CONWAY 2015: EDENHOFER - JAKOB 2017, 105-109).

- The global mean temperature has risen by 5 degrees,
- the ice in the polar regions has melted,
- the glaciers in the Alps are melting rapidly (with consequences for the water balance)
- The droughts in Africa have increased and are making life impossible,
- Australia and large parts of South America are uninhabitable.
- There are flight movements,
- people are fighting for land and living space in the northern regions of the world,
- the desert regions and the desertification are increasing.

This fictional retrospective from the future is about an age that is increasingly refining its forecasting skills but lacking the strength to act. It is about empirical observations vs. market fundamentalism with individual freedoms. A fight against climate change, so the fear from this point of view, could mean the end of individual liberties. Resistance is determined by economic interests. Particular interests can be successfully enforced with references to general values.

Climate policy is rational due to uncertainties about future damage. Tax regulations are not the end of individual freedoms; on the contrary, they enable future generations to exercise their rights to freedom.

In democracies, the state of knowledge, a public debate, the necessity of the matter and a consensus of society as a whole apply to implementation.

Authoritarian systems regulate opinions and scientific findings. Ideologies determine behaviour.

5 Online - Climate MOOC course - 2017

This MOOC provides a comprehensive overview of climate change and climate protection.

The online video course with certification covers the following topics in six chapters/modules:

- Organisational matters
- What is climate?
- What are climate models?
- Climate past
- Consequences of climate change
- Climate change and society

Chapter 1: Organisational matters
>http://mooin.oncampus.de > Climate MOOC (17 September 2017)

Chapter 2: Climate system and climate change

1. What is climate?

The climate system consists of the subsystems land, ocean, atmosphere, biosphere and ice sphere, which react differently to external influences. The climate is described by mean values, extreme values and frequencies of weather parameters such as temperature and precipitation over several decades.

2. What is the natural greenhouse effect?

Natural greenhouse gases in the atmosphere include water vapour, carbon dioxide, methane and nitrous oxide, which retain some of the Earth's heat radiation. This leads to warming and only then to life-friendly conditions on earth.

3. What is the anthropogenic greenhouse effect?

Human activities cause additional emissions of greenhouse gases, such as the burning of fossil fuels in the areas of electricity generation, heating and transport, as well as changes to the land surface, for example through deforestation, agriculture and livestock farming. This intensifies the greenhouse effect and is the main cause of climate change in recent decades.

4. What does climate change mean?

External and internal fluctuations result in a greenhouse effect that causes climate changes outside the natural fluctuations. The global mean temperature has risen by 0.8 degrees Celsius and the ice melt and sea level by 20 cm. There are regional differences.

5. What can you do?

Mitigation requires renewable energies and energy efficiency improvements.

Chapter 3: Models and scenarios

1. What are climate models?

Climate models depict the most important elements, processes and interactions in the climate system using mathematical equations based on basic physical laws.

2. What are climate scenarios?

These are not forecasts, but rather represent coherent and plausible images of the future. They describe future developments of the most important factors influencing the climate. Scenarios are tools that are used in politics and business when decisions need to be made. For the climate scenarios, the decision question is whether or not humanity will succeed in reducing greenhouse gas emissions in order to limit global warming to below 2 degrees Celsius.

3. What do predictions about developments in the 21st century depend on?

From the selected climate model, from the selected climate scenario taking into account the behaviour of society and from the decision of mankind for climate protection.

4. What climate changes do we expect in the 21st century?

An increased mean temperature of up to 5 degrees Celsius (compared to 1850) and the need to limit the temperature to below 2 degrees Celsius are to be expected.

5. Confidence in climate models and results

These are based on basic physical laws, the accuracy of measurements and anthropogenic greenhouse emissions.

Chapter 4: Climate of the past

1. Why is it important to study the development of the climate over the last 600,000 years?

By studying the climate of the past, especially the succession of ice ages and warm periods, we can further expand our understanding of the climate system - and thus also our confidence in climate models and forecasts for the development of the climate in the future.

2. How can we know how the climate has developed in the past?

We can obtain information from various climate archives. From sediment and ice cores, temperature, ice volume and CO2 content can be reconstructed over time; the C02 content in the atmosphere was lower in ice ages and higher in warm periods. Today, the CO2 content is significantly higher than these natural fluctuations.

3. How does the sequence of glacial and interglacial periods work?

The interval between ice ages is around 100,000 years. The succession of ice ages and warm periods is caused by fluctuations in the geometry of the earth's orbit. The C02 concentration in the atmosphere influences the dynamics of the change from cold to warm.

Sudden extreme changes in the climate system can occur at the transition from ice ages to warm periods. The collapse of ice sheets can disrupt the global system of ocean currents and thus influence the global climate.

4. Evidence of humanity through climate fluctuations

There has been another warm period for 100,000 years, despite climate fluctuations. There are collapses of advanced civilisations in connection with climate change.

5. Extraordinary climate developments since 1850

The temperature increase lies outside the natural fluctuations. The last ice age had about 3-5 degrees Celsius lower temperatures than in the Holocene. Without climate protection, the temperature could rise by 3-5 degrees Celsius.

Chapter 5: Consequences of climate change

1. What overall picture of the consequences of climate change is emerging?

Every increase in the global mean temperature increases the risks to humans caused by climate change. The risk of large-scale changes (tipping points in the climate system/ e.g. glacier and polar melting) also increases.

The effects of climate change are having very different impacts in different parts of the world and in different areas of life. A distinction is made between changes in mean values (e.g. temperature) and extreme events (e.g. more severe and more frequent floods or droughts).

2. What does risk mean?

The risk of climate change is made up of three factors.

- What climatic changes are happening?
- Are people directly or indirectly affected?
- How vulnerable are societies and what options do they have to adapt?

3. What are the effects in the oceans?

Oceans absorb around a quarter of anthropogenic C02 emissions. This causes their pH value to fall. This change is happening much faster than in the last 300 million years. Many calcifying marine organisms are being damaged as a result, with an increasing risk of negative impacts on biodiversity, fisheries (food security) and tourism.

4. Consequences for agriculture

Water shortages and soil erosion due to heavy rainfall and flooding as well as threats to yields in tropical regions can be observed.

5. Climate change with us

We are located in a climatic transition zone.

The rain shifts from summer to winter. There are dry periods on the one hand and floods, landslides and rockfalls on the other. There are increased health risks due to the hot days.

6. Cities in a changing climate

Cities and large urban areas are highly vulnerable due to their high population and value density. There are also localised effects such as wind attenuation and heat radiation. There are associated political and economic consequences. In any case, adaptation measures for people are necessary.

Chapter 6: Climate change and society

1. What does climate change mean for society?

Climate change is just one of many topics that people are concerned with in their everyday lives. People, regions and countries are affected by climate change in different ways. They do not all have the same interests when it comes to climate protection. However, climate change affects everyone in the medium and long term, which is why the international community must find a solution.

2. How can we adapt to the consequences of climate change?

Adaptation can be achieved through technologies (e.g. dyke construction, protective structures), behavioural changes (e.g. health behaviour in hot weather), political and economic decisions (e.g. retention). Adaptation is intended to reduce vulnerability. However, as climate change progresses, adaptation reaches its limits.

3. Does climate change influence conflicts and migration?

Regional conflicts and migration from crisis regions have very different causes, and climate change can indirectly intensify conflicts, for example through droughts, water shortages, crop failures and an increase in food prices as a result. This can also influence migration flows.

4. Goals of international climate policy

The aim of the international climate negotiations ("Framework Convention on Climate Change") is to reduce global greenhouse gas emissions (mitigation) and to tackle unavoidable climate change through adaptation measures.

A climate protection agreement has been under negotiation since 1992. So far, there has been no reduction in greenhouse gas emissions. Regional and national agreements complement the efforts. In 2010, it was decided that the increase in the global mean temperature should not exceed 2 degrees Celsius. 2 degrees Celsius is a political target based on scientific findings.

5. What does mitigation mean?

This refers to measures to limit climate change (reduction of emissions). In order to limit global warming to below 2 degrees Celsius, greenhouse gases must be reduced by 80 - 95 per cent by 2050.

Economists therefore propose setting up a global system of tradable emission rights and measures to promote new technologies and efficiency standards for buildings. Reducing emissions to the 2-degree path is entirely possible and compatible with further growth in economic prosperity.

Reflection

Dealing with this topic indicates the complexity of the existing problems.

The episode is about

- the necessary anticipation, consequences and
- Fields of action.

In terms of political education, climate policy is the most important factor.

- Those interested in education see an online course as a

significant contribution to further training.

- For adult education, the topic is a topical contribution to the future organisation of life.
- You realise the scope of the subject matter and the need for knowledge for future tasks.

The change in the earth is reflected in the diversity of the topics. This is reflected in the observation of a wide variety of factors (cf. PLÖGER-BÖTTCHER 2016, 175).

- In 2009/2010, glaciers worldwide retreated by 81 per cent.
- In 2012, the Arctic ice was 3.1 million km2 smaller than the long-term average.
- With a pH value of 8.06, the oceans are more acidic than they have been in the last 800,000 years.
- The heat records in the last 15 years have doubled compared to the previous 15 years.
- The Antarctic lost an average of 71 billion tonnes of ice annually between 1992 and 2011.
- Global water temperatures rose by 0.4 degrees between 1970 and 2010.

In addition to C02 pollution, there are other factors that increase global warming, such as deforestation.

The contribution has climate policy as a key sub-area, i.e. a subject area that requires action orientation.

Knowledge is also an essential basis for future decisions.

erwachsenenbildung.at
Wissen vernetzt. Bildung wirkt.

TEILNAHMEBESTÄTIGUNG

Hiermit bestätigen wir, dass Dr. Günther Dichatschek an der Online-Veranstaltung

wEBtalk: Klimaschutz in der Programmplanung umsetzen

am 14.09.2023 im Umfang von 1,5 Stunden

teilgenommen hat.

Inhalte

- Herausforderungen und Erfolgsfaktoren der Programmplanung zum Thema Klimaschutz
- Geeignete Kooperationen für die Umsetzung von Programmen zum Thema Klimaschutz

Folgende Lernergebnisse konnten erzielt werden:

- Die Teilnehmer*innen kennen Herausforderungen und Gelingensbedingungen für die Programmplanung zum Thema Klimaschutz und konnten eigene Fragen und Erfahrungen dazu im Kontext der Diskussion reflektieren
- Die Teilnehmer*innen haben erfahren, welche Kooperationen für die Programmplanung zum Thema Klimaschutz hilfreich sein können und konnten dies für den eigenen Anwendungszusammenhang reflektieren.

wEBtalks sind ein interaktives Online-Format von erwachsenenbildung.at. ExpertInnen teilen ihr Wissen und diskutieren ihre Erfahrungen. Das Online-Publikum ist eingeladen, Fragen zu stellen und mitzudiskutieren. Die Teilnahme am wEBtalk ist kostenlos und steht allen Interessierten offen.

Graz, 14.09.2023

Mag. Wilfried Frei
Geschäftsführung

Mag.ª Dr.ⁱⁿ Birgit Aschemann
Bereichsleitung Digitale Professionalisierung

Diese Online-Weiterbildung wurde veranstaltet von CONEDU.

Die Organisationseinheit DigiProf – Digitale Professionalisierung in der Erwachsenenbildung des Vereins CONEDU ist mit dem Qualitätssiegel der oberösterreichischen Erwachsenen- und Weiterbildungseinrichtungen (EBQ) zertifiziert und ist Ö-Cert Qualitätsanbieter.

Verein CONEDU
ZVR: 167333476 | Keplerstraße 105/3/5
A-8020 Graz | www.conedu.com
verein@conedu.com

CONEDU wird gefördert aus Mitteln des BMBWF und ESF.

Bundesministerium Bildung, Wissenschaft und Forschung

Bibliography I

The titles that are used and/or directly quoted in the article are listed.

Amery C.-Scheer H. (2001): Climate change. From fossil to solar culture, Munich

DichatschekG. (2017a): Didactics of Political Education, Saarbrücken

DichatschekG. (2017b): Interculturality, Saarbrücken

DichatschekG. (2017c): Adult continuing education, Saarbrücken

Eichhorst U.-Madry Th. (2013): Die Kehrseite der Medaille-Anpassung an den Klimawandel, Dossier of the Federal Agency for Civic Education, Bonn > http://www.bpb.de/gesellschaft/umwelt/klimawandel/38484/anpassung-an- den-klimawandel (as at 25.9.2013)

Edenhofer O.-Jakob M. (2017): Climate policy. Goals, conflicts, solutions, Federal Agency for Civic Education, publication series vol. 10163, Bonn

Feulner G.- Lesch H. (2010): The big book on climate, Cologne

Greenpeace in Central and Eastern Europe (2019): The climate crisis in Austria. The provincial report on the consequences of global warming by Greepeace Austria, Vienna > Global warming hits Austria hard > https://orf.at/stories/3136743/ (11 September 2019); Tyrol > https://tirol.orf.at/stories/3012231/ (11 September 2019)

Grefe Chr. (2016): Global Gardening. Bioökonomie - Neuer Raubbau oder Wirtschaftsform der Zukunft?, Federal Agency for Civic Education, publication series 1737, Bonn

Hantel M.-Haimberger L. (2016): Basic Climate Course, Berlin-Heidelberg

Madry Th.-Fischer J. (2014): Vorhersagbarkeit und Auswirkungen des Klimawandels, Dossier of the Federal Agency for Civic Education, Bonn > http://www.bpb.de/gesellschaft/umwelt/klimawandel/183026/auswirkungen- des-klimawandels (as at 17 April 2014)

Oreskes N.-Conway E.M. (2015): The End of the World: Chronicle of a Foretold Downfall, Munich

ORF - Global agriculture (2019): The vicious circle of global warming > https://orf.at/stories/3132807/ (8 August 2019)

Österreicher H. (2008): Natur- und Umweltpädagogik - Für sozialpädagogische Berufe, Cologne

Plöger S.-Böttcher Fr. (2016): Climate facts, Federal Agency for Civic Education, publication series 1734, Bonn

Rahmstorf St.- Schellnhuber H.J. (2020): Climate change, Federal Agency for Civic Education, publication series vol. 10520, Bonn

Stern N. (2007): The Economics of Climate Change. The Stern Review, Cambridge University Press

WBGU (2007): World in Transition: Climate Change as a Security Risk. Main Report 2006 - German Advisory Council on Global Change, Berlin-Heidelberg

Weingärtner L. (2014): Hunger and malnutrition, dossier by the Federal Agency for Civic Education, Bonn > http://www.bpb.de/internationales/welternaehrung/178484/hunger-und- fehlernaehrung (as at 12 June 2014)

II Alps

Introduction

An ancient cultural area, populated by around 14 million people, is changing due to climate change and tourism (cf. KASTLER 2018, 19)

For civic education, the change in a cultural area means a challenge to deal with aspects and developments in the Alpine region in an interdisciplinary way (cf. DICHATSCHEK2018).

Advertising, postcards, magazines, films and illustrated books show the Alpine region as an unspoilt landscape, a lovely alpine region with the romance of mountain huts, cliffs and peaks at sunrise and sunset, snow-covered pistes in winter and impressive cable car technology to make the mountains accessible.

- The people in the Alpine region are portrayed as being in a good mood and always have time for their guests, who are relaxing and tanned.
- Mountain hiking, ski slopes in the valley and training for many sports in the mountain landscape complete a picture characterised by climate change and tourism.
- Competition in the depiction of holiday dreams only exists in the triad of sandy beaches, the sea and palm trees.

However, there are also excerpts that show more than idyllic scenery and provide information on how a cultural and living space that is thousands of years old can be preserved (cf. BÄTZING 2018).

- As one of the largest mountain ranges in the world, the Alpine landscape stretches over 1200 kilometres from Nice to Vienna. The Tyrol region lies to a large extent in an alpine landscape.
- It is characterised by Mont Blanc, the highest mountain at 4807 metres, a living space of around 14 million people and a share of eight countries in the Alps with Monaco, France, Italy, Switzerland, Liechtenstein, Germany, Austria and Slovenia.

Throughout history, people have profoundly changed and reshaped the landscape ecologically, influenced it culturally and developed it socio-economically in order to secure their livelihoods.

The starting point for the considerations is the

- Completion of the university courses for political education and intercultural competence/ University of Salzburg,
- Completion of the Continuing Education Academy Austria and the Distance Learning Programme for Adult Education and Sustainable Development/ Evangelische Arbeitsstelle Erwachsenenbildung at the Comenius-Institut Münster and
- Examination of the specialised literature and the concrete changes in the Alpine region on site.
- Interdisciplinarity is the hallmark of a thematic debate - cultural geography, ethnology, political education and cultural studies (see HAID 2005/2008, BÄTZING 2015/2018, LUGER-REST 2018, MATHIEU 2018, DICHATSCHEK2018).

The article is divided into four subject areas in the context of civic education.

- Alpine geological history,
- Settlement and utilisation,
- Mass tourism and
- Alpine Convention.
- A reflection concludes the article with questions for the future.

1 Alpine geological history

With its geographically diverse landscapes, spaces, regions and cultural diversity, the Alpine region generates those longings that are addressed in advertising and that result in the use of leisure time, economic utilisation and personal living space.

Before the cultural-geographical conditions are discussed in more detail, the Alpine region should first be considered in terms of its geological and settlement history.

The Alps are a young mountain range in geological terms (cf. BÄTZING 2018, 3235). They were formed when Africa and Europe collided around 25 million years ago.

- The Adriatic Plate - today parts of Italy, the Balkans, the Alpine region and the Adriatic Sea - was still part of Africa and was driven like a wedge into the southern coast of Europe over millions of years.
- The collision caused the edges of the African and European plates to fold open where the Alps are today. The rock masses piled up on top of each other and are still growing today.
- Today, the African Plate is moving northwards at a rate of five centimetres per year.
- The Alps therefore grow by a few millimetres every year.
- Snow, water, wind and ice erode and reshape the mountain range.

2 Settlement and utilisation of the Alpine region

Colonisation led to cultural and economic changes that resulted in far-reaching ecological and socio-economic changes and transformations for the purpose of shaping and securing habitats.

- Clearances and ban forests resulted in small-scale settlements and economic areas with farming family structures that thought and planned in generations (cf. agriculture in the Alpine region BÄTZING 2018, 140-145).
- Today, such thinking and action is recommended against the destruction of valuable habitats.

- Trade and industry are often overlooked, with the transition to a service society taking place throughout Europe and the entire Alpine region (cf. BÄTZING 2018, 146-149).

2.1 Rural cultural landscape

T he conversion to a cultivated farming landscape has a socio-economic significance for the mountain farmers as well as a biological significance for the diversity of plant and animal species (cf. HAID 2005, VIDEK n.d.). The Alpine region is one of the biodiversity centres of Europe.

- Although the Alpine region only covers two per cent of Europe, it is home to around 40 per cent of the European plant world with around 400 plant species (cf. BÄTZING 2018, 62-65).
- Forests in the Alps are of particular importance (cf. closed forest areas, protected forest, forest pasture).
- The Alpine region is also considered a zone for water reserves (cf. BÄTZING 2018, 58-61, 164-167).
- There are at least 30,000 animal species in the Alps.
- Climate change is particularly affecting this diversity.

2.2 Alpine pasture farming

The Alpine region is considered an ideal area for mountain pasture farming (see MAIR 2019, 5; Climate change affects mountain pastures > https://tirol.orf.at/news/stories/2976847 [20.4.20019], "Dokumentation am Feiertag" Kuh, Schaf, Wolf & Klima > https://tirol.orf.at/tv/stories/3168228/ 13.8. 2022]).

- The different utilisation of low, medium and high mountain pastures is essential for mountain farming.
- Equally important are the various forms of cattle, bull, horse, sheep and goat pastures.
- Ownership structures such as private, co-operative, servitude and communal pastures are essential.
- The construction methods distinguish between alpine huts built of stone and wood.

- Food, clothing, costume, remuneration and alpine life are essential for the alpine pasture staff.
- Since their economic utilisation, the alpine pastures have also been considered a cultural area (cf. festivals, songs and alpine legends - recreational area).

In the Stella project, the University of Innsbruck is researching hydrological conditions in the Tyrolean Brixental Valley in the alpine pasture area (see https://www.uibk.ac.at/geographie/stella/stella-executive-summary.pdf > Version 2/6.11.2017 [21.12.2018]).

Different aspects of utilisation result from the conflicting interests of economy and ecology.

2.3 Alpine region as a source of raw materials

The Alpine region was an early area for coveted raw material sources, such as gold, silver, copper, iron and salt (cf. BÄTZING 2015).

- The mining areas became trading hubs and subsequently developed into markets and towns (cf. BÄTZING 2018, 116-123).
- With increasing colonisation came the development of
    - Migration of necessary labour and
    - cultural and religious diversity (think of the Reformation period with the spread of the Reformation faith through immigration in the mining regions/ example Schwaz; see BÄTZING 2018, 124-127).
    - Economic concentration and the political power that came with it resulted from the profits from the utilisation of mining production.

2.4 Industrial revolution

With the beginning of the industrial revolution, the cities began to dominate the countryside and the Alpine region.

- Small businesses lost importance.
- The utilisation of the large Alpine valleys began.

- The car, the lorry and the new railway lines, with tunnels in the mountains, subsequently gained in importance and became increasingly necessary.

Today, the major transversal routes run from north to south through the Alpine region and cause ecological problems, the consequences of which have a negative impact on the habitat and living conditions of the population (cf. the transport development of the Alpine region BÄTZING 2018, 132-139).

3 Mass tourism vs. soft tourism

3.1 Development of the alpine region

The development of the Alpine region was the basis for the mass tourism that developed at the beginning of the last century (cf. BÄTZING 2018, 150-159; GRUPP 2008).

In many valleys, tourism was the basis for colonisation and economic benefit.

Alpine urban centres and tourist centres were and are used intensively at this stage of development (cf. BÄTZING 2018, 160-163).

There are problem areas when tourism encroaches into areas that are not suitable for settlement.

Questions arise in the organisation of tourism with millions of guests when adventure landscapes are demanded and planned in a sensitive area.

The benefits for the majority of the population are questionable (cf. oversized infrastructure, high electricity and water consumption, high waste capacities, expensive building plots, traffic chaos)

Control mechanisms are required in order to be able to harmonise spatial development plans and regional developments.

IT notes

Alpinism > https://www.bergwelten.com/a/vom-ursprung-des-bergsteigens (11.12.2022)

The Conquest of the Alps > https://tvthek.orf.at/profile/zeit-history/13606695/zeit-geschichte-Die-Eroberung-der-Alpen-

Geschichte- des-Alpinismus-in-Qesterreich/14160039 (11.12.2022)

3.2 Leisure behaviour

The increasing leisure behaviour of the population in the Alpine region should be seen in this context (cf. BÄTZING 2018, 24-27).

- The urge to enjoy the alpine landscape is causing unrest for the game - there are around 200,000 animals in Tyrol alone - which means that the necessary annual shooting numbers cannot be met.

Functioning control mechanisms between the hunting community and tourism associations are required (cf. https://tirol.orf.at/news/sto ries/2953841/ [18 December 2018]).

- Climate change is increasing the pressure on winter tourism.
- Areas that were previously areas of retreat are to be opened up (see Freizeitparks im Hochgebirge BÄTZING 2018, 194-205).

Today, the Alpine region is a destination for around 120 million guests.

Efforts to preserve nature conservation as the preservation of reality in the Alpine region must be taken into account (cf. BÄTZING 2018, 168-169). The preservation and promotion of national parks with nature observations, an increase in biological knowledge and the preservation of large protected areas deserves more attention.

3.3 Alpine clubs

The national and international alpine clubs with around 2 million members maintain most of the trails and huts under increasingly difficult conditions.

- They are dependent on public subsidies and voluntary commitment ("volunteering").
- This forms the basis for Alpine tourism, which is gentle tourism.
- The commitment is also considered one of the major community projects.
- Beware of overcrowded mountain huts run by members,

dangers in the high mountains are often underestimated.

Association of Austrian Alpine Clubs (VAVÖ)

Alpine Club South Tyrol (AVS)

Alpine Association of Slovenia

Club4000

Club Alpin français

ClubAlpino Italiano (CAI)

German Alpine Club (DAV)

Austrian Alpine Club (ÖAV)

Naturefriends Germany

Naturefriends Austria

Naturefriends Switzerland

Austrian Mountain Guide Association

Austrian Tourist Club

Swiss Alpine Club

4 Alpine Convention 1989

The convention, which is binding under international law and was concluded in Berchtesgaden in 1989 by the environment ministers of Germany, France, Italy, Slovenia, Liechtenstein, Austria, Switzerland and the European Community, comprises a preamble and 14 articles.

4.1 Preamble

The preamble emphasises the importance of the Alps as a large interconnected natural area in Europe,

- the specific and diverse nature, culture and history as an excellent living, economic, cultural and recreational area in the heart of Europe, in which numerous peoples and countries participate.
- The importance for non-alpine areas, including as a carrier of

important transport routes, is recognised.

- The Alpine region is also emphasised as a refuge and habitat for endangered plant and animal species.
- The differences arc recorded
    - of the individual legal systems,
    - the natural conditions and colonisation,
    - agriculture, forestry and economic development,
    - the traffic load and
    - the type and intensity of tourist use.
- It recognises the growing pressure on the Alpine region, the threat to ecological functions and the high costs involved in remedying them, usually only in the long term.
- It expresses the conviction that economic interests must be harmonised with ecological requirements.

4.2 14 articles

Article 1 covers the scope of application.

Article 2 concerns general obligations such as the respect, preservation and promotion of the population and culture, spatial planning, air pollution control, soil protection, water management, nature conservation and landscape management, mountain agriculture, mountain forests, the harmonisation of tourism and leisure activities, traffic pollution, energy production and waste prevention.

Article 3 describes the research activities and systematic monitoring of the areas referred to in Article 2.

Article 4 documents co-operation in the legal, scientific, economic and technical fields.

Article 5 covers the organisation of the Conference of the Contracting Parties (Alpine Conference).

Article 6 covers the remit of the Alpine Conference.

Article 7 deals with decision-making in the Alpine Conference.

Article 8 refers to the provisions of Article 7 with the tasks of the Standing Committee.

Article 9 refers to the Secretariat of the Alpine Conference.

Article 10 concerns amendments to the Convention.

Article 11 refers to protocols and their amendment.

Article 12 regulates the signing and ratification of the Convention.

Article 13 regulates the cancellation of the agreement.

Article 14 regulates the notifications.

IT note

Current version 2021 > http://www.ris.bka.gv.at/GeltendeFassung.wxe?Abfrage=Bundesnormen&Gesetzesnummer=10010876 (22.9.21)

5 Reflection

5.1 The future of the Alpine region

The question of the future of the Alpine region is a key issue for political education.

"If the development of the Alps continues as before, the last remaining cultural landscapes will gradually disappear in the foreseeable future and the Alps will consist exclusively of forested and urbanised regions" (BÄTZING 2018, 210).

5.2 Objectives

Based on HAID (2005) and BÄTZING (2018), the objectives would be

- an upgrading as a decentralised living and economic area,
- in connection with the modernisation of Europe, the preservation of the Alpine region as an open space with independent development,
- This means utilising valuable resources (such as quality products/livestock farming typical of the region, arable and fruit farming, wood, energy use),
- Conversion of tourism to utilise the regional economy (see

Alpine tourism/evaluation and change > http://rn.bpb.de/apuz/25886/alpentourismus-bewertung-und-wandel?p=all [2.2.2019]),

- faster Internet with decentralised workplaces,
- new cultural landscapes with habitat protection. This requires a specific infrastructure and support, supervision and counselling for community projects.
    - Expansion of school and extracurricular educational facilities,
    - Provision of medical facilities and patient care,
    - Social care for young people and senior citizens,
    - Adequate connections to public transport and a suitable transport network,
    - sufficient administrative units,
    - Possibilities of workspaces,
    - Support and promotion of mountain farming,
    - Creation of leisure spaces.
- Greater attention must be paid to the importance of the Alpine Convention in safeguarding European interests in the Alps (cf. transit traffic, water use, recreational areas/protected areas - nature conservation national parks).

5.3 Fields of activity of political education

The areas of responsibility are

- an agricultural policy with the aspect of preserving mountain farming - economically, ecologically, socially and culturally,
- climate change and climate protection as a complex issue of environmental policy and
- increased attention to the sensitive Alpine region in the context of global considerations and global learning.

A conflict of interest arises from the bipolarity of economy and ecology.

Political education highlights the controversy and seeks consensual solutions in individual cases.

The beauty of the Alpine landscape and its uniqueness - both in and out of school - should be increasingly emphasised in the context of political education in

- projects,
- Excursions,
- Explorations,
- Images and
- practical experience (cf. the dissertation on the culture of the mountain guide profession [Switzerland] by HUNGERBÜHLER 2014).

Increasingly, we are seeing the consequences of climate change with a

- Warming in the high mountains (cf. rising snow line, glacier melt, lower water runoff, permafrost melt, severe weather and natural hazards [heavy rain, mudslides and rock avalanches] as well as increasing areas of drought),
- Shifting the flora and fauna and

- Appearance of new pests.

5.4 IT - Notes

University of Innsbruck - "International Mountain Conference 2019'78-12, 2019

>https://tirol.orf.at/stories/3011876/ (8 SEPTEMBER 2019)

ORF.at - Climate change is causing the Alps to crumble faster > https://salzburg.orf.at/stories/3014758/ (27.9.2019)

Bibliography II

The titles that are used and/or directly quoted in the article are listed.

Alpine Convention (2013): Sustainable Tourism in the Alps, 4th Report on the State of the Alps, Innsbruck

Bätzing W. (2015): The Alps - History and Future of a European

Cultural Landscape, Munich

Bätzing W. (2018): The Alps. The disappearance of a cultural landscape, Darmstadt

Dichatschek G. (2018): Didactics of civic education. Theory, practice and fields of action in the didactics of civic education, Saarbrücken

Grupp P. (2008): Fascination of the mountain. The history of alpinism, Vienna

Haid H. (1986): Vom alten Leben. Passing forms of existence and labour in the Alps, Rosenheim

Haid H. (2005): New Life in the Alps. Initiatives, models and projects in organic farming, Vienna - Cologne - Weimar

Heinich - Böll - Stiftung/ Bund für Umwelt und Naturschutz Deutschland/ Le Monde Diplomatique (2019): Agricultural Atlas 2019: Facts and figures on EU agriculture, Berlin

Hungerbühler A. (2014): Kings of the Alps. On the culture of the mountain guide profession, Bielefeld

Kastler U. (2018): There is no ideal world in the Alps, in: Salzburger Nachrichten, 13 December 2018, 19

Luger K. - Rest Fr. (eds.) (2018): Travelling in the Alps. Experience, Spatial Transformation, Imigation, Innsbruck-Vienna-Bolzano

Mair B. (2019): Between home and recreation, in: Tiroler Tageszeitung, 15 April 2019, 5

Mathieu J. (2015): The Alps. Space - Culture - History, Stuttgart

Pfeiffenberger A. (2019): The risk on the mountain is underestimated, in: Salzburger Nachrichten, 8 January 2019, 1

Vitek E. (n.d.): The flora of the Austrian Alps, Natural History Museum, Vienna

14 Alpine ethics

In the following, an attempt is made to present Alpine ethics as a practical form of application in everyday life and, if necessary, to put it up for discussion as a model for the future of the Alps as a living space. In the following, an attempt is made to formulate a kind of decalogue for Alpine ethical issues.

14.1 Introduction

Friends of the Alps, alpinists, mountain hikers, sports enthusiasts and nature lovers are committed to the preservation of mountain landscapes.

Alpine clubs and associations, which are open to all those who agree with the objectives, are generally regarded as contact points for competent and reliable partners. There are also recognised experts among state-certified mountain guides and army mountain guides, alpine botanists and zoologists, environmental experts and mountain agriculture experts, folklorists and human geographers.

An Alpine partnership - with or without associations - sees itself as independent of political parties and denominations.

14.2 Youth - Families - Seniors

Active and shared mountain hiking and mountaineering as well as experiencing nature is a central concern for all age groups.

What is needed is an overall social commitment from the various socio-political groups and interest groups. Bringing them together in this way and using cooperation models will promote mutual cross-generational appreciation.

14.3 Classic forms of alpine activities

Classic mountaineering includes mountain hiking and challenging mountain tours.

- The recreational effect for body, soul and spirit should find its place here, as should sporting activity.
- The starting points are the principles listed above and below, in particular personal responsibility and knowledge of alpine ethical standards.

14.4 New sports - trends - developments

New alpine sports, trends and developments should be compatible with nature and the landscape. Alpine ethical principles should be complied with. The focus should be on promoting popular sports and supporting competitive sports.

14.5 Training for the alpine region

Ongoing training and further education for managers in the various areas of activity seems essential. The target groups are also the members of the alpine clubs, beginners and advanced skiers of all ages, volunteers and full-time employees.

General adult education is also becoming increasingly important for cultural work in the Alpine region. Other topics include aspects of nature and the environment, safety and mutual tolerance and interculturality.

14.6 Nature and environment

The preservation of nature and the environment is a central concern of alpine ethical principles.

- A nature and environmentally friendly approach to the mountain landscape and its sensitive flora and fauna requires education and training.
- Democratic political influence is also required (cf. the importance of civic education with its thematic areas, in particular environmental education and climate change).

14.7 Alpine infrastructures

Trails, huts and ascent aids and their maintenance play a major role in the success, importance and preservation of the Alpine region.

This also includes Alpine tourism.

For active leisure activities, the need for and construction of appropriate facilities for alpine training must be taken into account in terms of spatial policy and environmental compatibility.

14.8 Security

The factor of safety and mountain rescue is of the utmost importance when carrying out activities in the Alps. Prioritising accident prevention and supporting measures in mountain rescue are part of the Alpine ethic.

14.9 Culture-Art-Tradition-Customs

Mountain regions in the Alps have various cultural areas with historical significance. The cultivation and promotion of Alpine culture and art, languages, traditions and customs therefore belong to the subject areas of folklore/ethnology, interculturality, human geography and political education.

14.10 Information and public relations work

There is a need for open, objective and timely information and public relations work, both within and outside the Alpine region. Alpine information is increasingly becoming a focal point.

III Sustainability

1 Introduction

The starting point of the dimensions and concept for the author are the

- Completion of the distance learning programme "Sustainable Development" - Evangelische Arbeitsstelle Fernstudium im Comenius - Institut Münster (2020),
- Completion of the 10th University Programme in Political Education - University of Salzburg - Klagenfurt (2008) and
- The company's involvement with specialist literature and the increasing topicality of the subject.

As an introduction to the topic, "sustainability" in everyday language means something like lasting, persistent or having an effect over a longer period of time.

The term "sustainable development" is used to describe how humanity should develop, emphasising the processual nature of natural and social change (see FERNKURS NACHHALTIGE ENTWICKLUNG EINFÜHRUNGSHEFT 2014, 12).

1.1 Sustainable development

The normative core of "sustainable development" is the idea of inter- and intragenerational justice in view of dwindling natural resources and a declining ability of ecosystems to absorb emissions and increasing environmental degradation.

The concept commits to ensuring that people can lead a safe and autonomous life. The starting point is to secure and, at best, restore the basis of life for everyone today and in the future (cf. MEISCH 2014, 6).

It would be misleading to speak of a precisely definable state of "sustainable development", especially as there are different organisational spaces depending on culture, place and time.

The reasons for this are the physical and socio-cultural conditions in their diversity (cf. ecosystems, societies and cultural identities).

In addition, the change in conditions over time due to natural and social processes as well as ethical ideas to be examined (cf. climate change; demography, socio-political upheavals; justice) must be taken into account.

The transformation to a more sustainable society is a knowledge-based and ethically orientated process of searching, learning and experiencing (see GRUNWALD - KOPFMÜLLER 2012, 12).

Science contributes to these processes through certain forms of knowledge.

1.2 Transformation to a sustainable society

Participatory and transformational education enables active participation (cf. the endeavours of educational science and didactics of civic education; DE HAAN 2004, 39-46; DICHATSCHEK 2017a, 2017b).

The contribution is intended to encourage "Education for Sustainable Development" (ESD) in all areas of education in the context of civic education.

- The programme is aimed at schools, teacher training and adult education.
- School pedagogy, university didactics and adult education are required.

In the following, the combination of economy-ecology-social issues / "three pillar model", health-sport-leisure, science - sustainability and non-sustainability or risk policy, culture and a project design are presented.

The article deals in detail with sustainable mobility and, as an element of a pedagogical development process, factors in the learning culture of continuing education.

The study does not claim to be exhaustive; it focusses on personal views of the authors.

The aim is to contribute to an objective discourse in political education.

2 "Three-pillar model" Economy, ecology and social issues

The three pillars were developed in the 1990s and were used as a benchmark for sustainability at the World Summit in Johannesburg in 2002.

The balance and equal importance of the three areas are central to sustainable development. The three goals are mutually dependent. In this way, the economic, ecological and social performance of a society should be ensured and improved.

2.1 Economic sustainability

Good management is required.

Sustainable companies must also make a profit. Investments should be made in technical innovations, new employees and training programmes. However, profit maximisation should not be the only goal. Sustainable companies can also pursue new goals, such as improving the quality of life or promoting environmental protection, health promotion, family friendliness, training and further education projects.

On a national level, economic sustainability means keeping public debt low so that future generations are only burdened to a small extent. External economic balance must be maintained so as not to penalise other countries. Export-strong countries make other countries dependent on exports. These countries are unable to develop a locally competitive economy, which results in high unemployment. The country's indebtedness should be prevented.

2.2 Ecological sustainability

It calls for the environment and natural resources to be conserved. Companies and countries should work towards the conscious use of water, energy and finite raw materials.

The pillar also stands for conscious consumption of the earth's natural resources and how renewable resources can be replaced. The aim is to avoid damage to the ecosystem and subsequently promote biodiversity. This also means keeping emissions as low as possible.

The pillar also stands for a conscious approach to human health, i.e.

prevention and active health promotion.

2.3 Social sustainability

What is required is the position of people at the centre, their dignity and free development. This basic assumption represents the minimum.

Exploitation, forced labour and child labour are therefore contradictory.

The "Social" pillar calls for fair pay, the realisation of employee interests and personal development in terms of lifestyle, educational measures and professional activities.

Orientation towards the common good in society is the socio-political goal.

2.4 Criticism

A whole range of alternative sustainability concepts, which are significantly more complicated, are also being criticised.

One major criticism is that it is difficult to apply. The model provides guidelines, but no concrete solutions. For this reason, it is probably applicable to many areas, but its openness is criticised.

The pillars are said to be open to almost all economic and social policy objectives such as economic growth, full employment or an increase in tax revenue. In this way, the model greatly expands the concept of sustainability.

It is also criticised that the three pillars are theoretically but not actually of equal importance. The fact is that the ecological pillar is subordinate in practice in modern society ("weak sustainability"). Natural resources are balanced out by human and physical capital; the only thing that counts is prosperity, even at the expense of nature.

"Strong sustainability" should be the goal of ecological sustainability with the protection of natural living conditions as a basic prerequisite for ecological and social stability.

The model can also be found as a concept in Art. 1 of the Treaty of the European Community.

There are calls for the expansion of a fourth pillar with a political-institutional dimension.

In 2015, the UN adopted the "Agenda 2030" for sustainable development. Over 13,000 companies have signed up to the UN Global Compact. They are committed to universal principles in the areas of human rights, labour standards, the environment and corruption prevention. These areas can be traced back to the model.

Evangelische
Arbeitsstelle
Fernstudium

Zertifikat

für

Herrn
Dr. Günther Dichatschek MSc

über den erfolgreichen Abschluss des Fernkurses

Nachhaltige Entwicklung

der Evangelischen Arbeitsstelle Fernstudium im Comenius-Institut e.V.

STAATLICHE ZENTRALSTELLE FÜR FERNUNTERRICHT
ZUGELASSEN
ZFU

Münster, 17. August 2020

Frau Dr. Ada Gertrud Wolf
Leiterin der
Evangelischen Arbeitsstelle Fernstudium
im Comenius-Institut e.V.

3 Health, sport and leisure

3.1 Healthcare

In Austria, the Federal Health Commission and the Council of Ministers adopted targets in 2011 for the sustainable improvement and maintenance of health in conjunction with a reduction in the burden on healthcare (see FEDERAL MINISTRY OF HEALTH AND WOMEN 2012).

Based on an efficient healthcare system, Austria's life expectancy is below the EU average. It is therefore a sustainable goal to increase the number of healthy years. The interdependence of health, sporting activities and the utilisation of relaxing and meaningful leisure time is a given.

In terms of "Education for Sustainable Development/ ESD", the goals influence the factors of education, labour situation, social security and environmental influences.

3.2 Basic principles

Basic principles indicate the dimensions.

- Alignment with the influencing factors,
- Consideration of different policy areas,
- Promotion of equal opportunities,
- Orientation towards resources,
- Orientation towards a benefit,
- Sustainability as a long-term benefit,
- Impact orientation as relevance and
- Comprehensibility for non-specialists.

Particular importance is attached to the first three and sustainability.

4 Science

4.1 Claim

The claim of science is

- the realisation of theoretical and
- subsequently realisable practical results and solutions in the diversity of their disciplines, such as the
    - Natural sciences,
    - Humanities,
    - Engineering sciences,
    - Economics,
    - Social sciences,
    - Cultural studies,
    - Educational sciences and
    - respective ethical approaches (cf. generally analytical ethics, applied ethics).

4.2 Boundaries

Limits and shortcomings arise from the application of knowledge, the conception of science, the plurality of forms of knowledge and the shattering of scientific optimism.

Science is without doubt the most important source of information about opportunities and risks.

The task of ethics is to bring in the results on the basis of standards of success.

5 Non - sustainability - risk policy

In the following, the lack of sustainability ("non-sustainability") and risk policy are treated as an expression of this dimension.

5.1 Non-sustainability

From the understanding of sustainability, the special nature of contemporary society can be characterised by the concept of a society of "non-sustainability" (cf. VOLKMER - WERNER 2020, 229-230).

- Wealthy consumer societies in the Global North are infected by the virus.
- The destructive force is significant in social, ecological, economic and democratic terms.
- Certain social groups, entire countries and also ecological and economic conditions are affected. Despite their systemic relevance, they are held in low esteem.
- Society is resolutely defending its freedoms, values and lifestyle, "whatever the cost".

The democratic and fourth dimension, particularly interesting for political education, has reacted with determination and great capacity for action in the first phase with a social alliance and solidarity in connection with systemically relevant groups.

- As a result, the measures were seen by protest movements as a threat to fundamental rights and the rule of law.
- In fact, the pandemic is accelerating changes that were already emerging (see "post-democratic turn") and supporting unsustainability.
- In recent decades, even in established democracies, there has been increasing talk of disenchantment with politics, disenchantment with political parties and a crisis of democracy.
- Reforms should overcome political phenomena.
- The increasing complexity of political processes, the internationalisation and plurality of modern societies appear to be overstretching liberal democracies.
- Increasing dysfunctionality relates to democratic institutions and procedures, their suitability and the realisation of modern demands for freedom and self-determination.
- This political process is leading to activities by many political actors such as decision-makers, liberal elites, climate activists, marginal minorities, national populists and other groups that are developing an ambivalent relationship with democracy (cf. HIRSCHMANN 2017; VOLKMER - WERNER

2020, 233).

- Accordingly, a transformation of democracy of increasingly actualised and marginalising conditions of freedom and self-realisation or personal autonomy could be expected.
- This crisis is expanding into the unsustainability of democracy.
- Liberal democracy also shares responsibility for the ecological and social unsustainability of the consumer society.
- In cultural terms, liberal democracy also appears to be emaciated.
- An unsettled political situation is emerging here with threats to liberal democracy in conflict situations with illiberal and authoritarian political phenomena and the need for sustainability (see SCHEIDIG 2016, KOST - MASSING - REISER 2020).

5.2 Risk policy

Behind risk management such as imminent damage, disasters and a state of emergency is a mode of state policy / "risk policy".

According to Ulrich BECK (1986, 1998) and Niklas LUHMANN (1991), there is a "risk society" with technical, economic, ecological and health risks. The endeavour to create a precautionary state is to be seen in a statehood.

5.2.1 Structural features

Structural features of a modern risk policy (cf. VOLKMER - WERNER 2020, 242-247)

- Risk policy is a policy of the negative - attempts are made to avoid or mitigate negative conditions. Intra-societal risks include economic crises, unemployment and crime, while extra-societal risks include technical risks and ecological hazards. The policy of the negative is combined with the policy of the positive, for example as a welfare state.
- Risk policy as security in the future - security at the expense of freedom and dynamics such as economic market structures and technical progress.
- Risk policy is dependent on scientific expertise - such as

economics and sociology, health science and medicine, economics, computer science, climate research and education science.

- Collective and individual risk management - individuals carry out preventive work themselves (personal safety, social precautions), the entire population minimises risks (crises, emergency plans).
- Risk policy with different risks - containment measures and consequential risks.
- Risk policy in permanent mode or crisis mode - ongoing risk policy or extensive use of funds.

5.2.2 Factors

The following factors reinforce risk structures such as

- globalised capitalism and highly complex financial markets,
- global politics with the loss of US hegemony, the rise of China, the strengthening of Russia, the development of "failed states" in the Middle East, for example, and internal tensions in the European Union,
- Digitalisation of the economy, education, administration, medicine, the military and private relationships as a function of highly complex technology,
- Globalisation easily leads to regional crises that can subsequently spread across regions, such as the financial crisis, coronavirus crisis and refugee crisis,
- Socio-cultural tensions have increased in Western societies since 2010 and have in some cases led to political polarisation (see RECKWITZ 2019),
- Risks have increased due to the election of national populist politicians, political protests (yellow waistcoats) and the spread of conspiracy myths,
- Climate change harbours a wealth of ecological risks (such as long-term warming, extreme weather, drought, loss of biodiversity).

The factors mentioned do not speak in favour of an exception to risk policy, but rather in favour of its importance in late modern politics.

One idea of resilience can be transferred to the collective level, as can that of vulnerability. A resilient state would be in permanent mode through risk policy. Climate change policy would be an example of this (cf. VOLKMER - WERNER 2020, 249).

6 Culture

The term culture refers to creative and artistic work. A certain way of life is referred to as sophistication. Certain habits are typical of specific areas, such as youth culture or student culture.

According to the Latin meaning "cultura" care, one speaks of cultivated landscapes or monoculture.

6.1 Cultural sector

According to UNESCO 2007, the cultural sector includes cultural and natural heritage, performances and festivals, visual arts, crafts and design, books and press, and audiovisual and digital media (cf. FREERICKS - HARTMANN - STECKER 2010, 197-198).

Sustainable development is a comprehensive challenge. In addition to the usual solutions, it is about cultural norms and behaviour in a modern society. The call for cultural change is becoming ever louder.

- The myth of growth and, as a result, the consumer culture must be tested for suitability.
- Helpfulness, co-operation, self-interest and public interest in the context of global justice must be brought into balance.
- In 2008, a country like Ecuador took notice by including the rights of nature in its constitution.
- What is needed are new guiding principles and people who are prepared to find new creative ways into the future.
- Breaking up old ways of thinking is the core business of culture and art.
- In the 1972 Club of Rome study "Limits to Growth", co-author Donella Meadows considered a change in the cultural norms of Western societies to be essential (cf. the 1987 Brundtland

Report, which also assumes a comprehensive process of change).

- The "UNESCO Declaration on Cultural Diversity" 2001 is based on the importance of the diversity of cultures and the integration of culture in sustainable development.

6.2 Values and attitudes

The question of values and attitudes must not be neglected in the context of civic education (cf. the educational mandate of the subject areas of ethics, religion, civic education and interculturality). It is about positive visions of the future and dimensions for a successful life.

- The project presented as an inspiration for larger projects offers the opportunity to take responsibility and experience a chance for personal development.
- Being an active part of society is worthwhile in terms of educational endeavours.
- As a result, it is natural to see cultural education in the context of sustainable development.

It is about strengthening the competences of participation - cooperation - responsibility - new ways of thinking - respect for nature and the diversity of cultures for a transition to a post-industrial society.

Internet reference

Leipprand Eva (2013/2012): Culture, Education and Sustainable Development, in: Kulturelle Bildung Online > https://kubi-online.de/artikel/kultur-bildung- nachhaltige-entwicklung (13.10.20)

7 Field of learning and action

7.1 "World Decade for Sustainable Development"

Based on the UN initiative of the "Brundtland Report" in 1987, this is an area of education that emerged in 1992 as a result of the Rio Conference.

- The UN member states have committed themselves to the guiding principle of sustainable development in their education systems.
- The "World Decade for Sustainable Development" was proclaimed in 2005-2014.

7.2 Subject areas

This results in complex subject areas in the context of pedagogy and specialised didactics in the

- secondary (school),
- tertiary (FH, university) and
- quaternary (adult education) education sector

and thus in school lessons, university teaching and courses.

Examples of these are

- Justice - intergenerational justice, gender justice,
- Peace - conflicts, migration, recognising the foreign,
- Future - food supply, energy resources, health, traffic/transport, economic models,
- Environment - climate protection, landscaping, biodiversity, mobility and traffic volume
- Sustainability - political-economic-ecological-social and cultural coexistence.

8 Project topics/selection

8.1 Suggestions for school development - aspects of a new school and learning culture

Personal development > Learners

Personnel development > Teachers

Education management > Education management

8.2 Political education

Austria

State structure

Parliamentarism, democracy, participation

Multilingualism, interculturality, transculturality

Europe

EU, Council of Europe, OSCE, OECD

Alpine habitat (Alpine Convention)

Populism

Dismantling democracy

8.3 Intercultural competence

Migration

cultural diversity and interculturality

global learning processes

8.4 Educational processes

Civic education - democracy education

Career guidance - Youth unemployment

Health education

Economic education

Ecology

Applied Ethics - Religion

Further training

8.5 Parental work

- Education - standards and values
- School - Teaching and learning development
- Education system - autonomisation and reform efforts
- Student and parent counselling and educational counselling - Student counselling and
- Adult counselling (cf. educational, family, career and legal counselling)

8.6 Didactics - Project work

8.6.1 Basic knowledge

- Teaching quality - teaching and learning culture
- Teaching and learning methods
- Specialised didactics

8.6.2 Forms of learning

- Teamwork
- Aspect investigations - expert discussions
- Case studies
- Workshops - Working groups
- problem-based learning,
- Learning in subjects or subject disciplines,
- interdisciplinary learning,
- critical learning and
- Net-based learning

8.6.3 Localisation of the project

- Networking and setting up your own homepage with support - example for school http://www.schulwiki.org/wiki.cgi?
- Documentation
- Presentation - Learners, teachers - Parent work
- Networking
- Press relations - Public relations
- Publication

Sustainable mobility - future viability

In the following, a project is presented for discussion in the context of the "VCÖ - Mobility Award 2022".

9 Subject areas

9.1 Passenger and freight transport

Passenger and freight transport is growing steadily in the forms of transit traffic and urban transport systems.

Mobility and accessibility are prerequisites for

- social participation,
- economic exchange,
- Employment and
- Prosperity.

At the same time, the current transport system is causing severe environmental pollution. The volume of traffic in Austria and its neighbouring regions has increased massively in recent decades. Especially passenger transport in addition to freight transport.

Even if vehicles have become cleaner and quieter, motorised traffic still causes a lot of pollution through

- emissions of greenhouse gases,
- air pollutants,
- Noise,
- Land utilisation and
- consumption of resources has a negative environmental impact.

9.2 Sustainable mobility transport

The question arises as to how the mobility of people and goods can be maintained and secured without traffic placing an excessive burden on people and the environment in the long term.

Sustainable mobility is therefore at the centre of considerations. With its "Strategy for a climate-neutral Europe", the European Commission has set itself the goal of decarbonising the European mobility system by 2050, i.e. making it greenhouse gas-neutral.

10 Political discourse

10.1 FieldsPolltlscherBlldung

The topic covers fields of political education.

The political decisions must be implemented with measures in order to achieve the goals.

A strategy in the area of transport performance comprises the four fields of

- Avoidance of traffic,
- Shift to more environmentally friendly modes of transport such as rail or ship,
- Increasing energy efficiency and
- use post-fossil greenhouse gas-neutral fuels and electricity.

In order to organise transport efficiently using alternative fuels and drives, it is necessary to reduce the energy demand of the transport sector. Avoiding and shifting traffic is therefore essential.

10.2 Sustainable transport

Avoidance in particular is often misunderstood.

The aim is not to prevent traffic,

- shorten transport routes by changing settlement structures and production locations or production structures, or
- increase the utilisation of vehicles.
- It's about more mobility with less traffic.

Sustainable transport requires a bundle of measures (intermodal transport), as individual measures cannot have the necessary effect.

- The vehicle and its drive technology must not take centre stage alone.
- Climate protection targets can only be achieved with an integrated approach with non-technical measures.
- What is needed are economic incentives with the aim of changing behaviour, and
- settlement and transport planning,

• traffic avoidance and environmentally friendly modes of transport.

11 Environmentally friendly everyday transport

Promotion of cycling and walking - cycle paths and cycle highways, parking spaces and signage - space for walking and more walking concepts

Bus and rail - expansion of regular services and more stops - direct and fast connections - improvement in service quality

Transport networks - car sharing through preferred parking spaces - combined use of different modes of transport - transparent tariff system

Private car traffic - expansion of parking management - standard speed in local traffic and motorways

12 Sustainable tourism

Following LUGER (2022, 254 - 297), there are many different ways to achieve sustainability in tourism. The starting point should be the type of transport and traffic to and from holiday destinations (cf. REIN - STRADAS 2015).

12.1 Travelling

Up to 80 percent of travel within Austria and from neighbouring countries to Austria's holiday resorts is by car (see Internet information on the consequences of climate change > https://ccca.ac.at/wissenstransfer/apcc/broschuere-der-oesterreichische- tourismus-im-klimawandel, 23.6.2022).

Individual car traffic is the trigger for "overtourism". It is about exceeding the carrying capacity of a destination, the excess in city centres, historical sites and excursion and recreation areas.

12.2 Alpine region

The Alpine region is also characterised as a sensitive natural, living and tourist area with a high level of automobile traffic.

- The provision of environmentally friendly means of transport is of central importance for a climate-friendly regional policy.

- A transport network of rail (train) and road (bus) makes it easier to switch to public transport. A good luggage service, shuttle services and multimodal offers are desirable.
- The importance of local railways and their connections should not be underestimated.
- Examples such as a travel platform for the Villach-Faker See tourism region have been recognised by the VCÖ, and digital travel maps facilitate travel habits (cf. LUGER 2022, 256). Note also the need for counselling, support and, if necessary, guidance (cf. the importance of professional development).

Of international importance is the European Union's "Green Deal" as a law to make Europe emission-free by 2050. Aly's interim target is to reduce greenhouse gas emissions by 55 per cent by 2030.

As in all sectors of the economy, tourism faces the following challenges

- Savings in the energy sector,
- Conversion of the heating processes and
- thermal insulation of the accommodation facilities.

In terms of visions and measures, the demands of the sustainability discourse in documents and the objectives and framework conditions for tourism in the EU's 5th Action Programme and the protocols of the Alpine Convention on Environmental Development have been included in Austria.

Internet reference

https://www.ris.bka.gv.at/GeltendeFassung.wxe?Abfrage=Bundesnormen&Gesetzesnummer=10010876 (20 JUNE 2022)

The principles and objectives of a sustainable tourism and leisure industry were defined as early as 1995, but implementation is lagging considerably behind, and not only in Austria (cf. LUGER 2022, 259).

The development to protect the Alpine region as a living and cultural space is in the interest of society as a whole and of the public, a branch of industry with tourism infrastructure, transport policy and spatial planning (cf. DOLLINGER2021).

Sustainable development in the Alpine Central European cultural area takes the following factors into account (cf. LUGER 2022, 261-262)

- Spatial planning,
- Traffic,
- Agriculture and forestry,
- Environmental and nature conservation and
- Water and energy supply.

12.3 Plan T - Masterplan for Austrian tourism

The 2019 masterplan is entitled "On the way to becoming a sustainable tourism destination" (see LUGER 2022, 262 - 268).

The guiding principle of sustainability is summarised below.

- Sustainability as a global responsibility and contribution to the realisation of the UN goals
- Intact nature - safeguarding the basis of life
- Tourism as harmony between people and nature - Resource efficiency
- Expansion of sustainable mobility and transport solutions
- Rethinking tourism - culture of cooperation - utilising digital potential
- Employee problems - attractiveness of tourism work
- Marketing - Management

Internet information

Plan T > https://info.bmlrt.gv.at/themen/tourismus/masterplan-tourismus- plan-t.html (25.6.2022)

Labour shortage > https://orf.at/stories/3268573/ (25.6.2022)

12.4 Ecotourism

A vision should ultimately be addressed with a quote from Confucius: "If you go to another country, you should know what is forbidden there" (cf. LUGER 2022, 271, 272-285).

- Do not destroy what is being sought when it is found.
- Equalising measures in tourist ghettos - adapting the infrastructure and improving the mobility structure.
- Steering measures to shift visitor flows - cooperation between cultural institutions and encounter programmes/destination management.
- Tourism and leisure industry - field of action of regional policy - sustainable tourism services
- Quality standards for nature-based tourism - regional reference - avoidance of damage and utilisation of natural resources

13 Transport policy for sustainable mobility

13.1 Fundamentals

The economic, environmental and transport factors pose the challenge of how the state, as a responsible institution with an interest in society as a whole, can shape the network of effects for future generations (cf. SCHWEDES-RAMMERT 2021, 312-319).

Future mobility must be guaranteed without being at the expense of people and nature. This is a challenge for planning science and political science. Pedagogically, the didactics of civic education and a learning culture of continuing education are responsible (cf. DICHATSCHEK2017a, 2018).

Planning science opens up a new dimension, different from traditional transport planning. The design of mobility requires adaptation to social and ecological demands.

Transport policy relating to mobility and its design is located in the political arena.

An examination of political science reveals the challenge of starting points for design and implementation.

The basis in a democracy is a regulatory policy of social and ecological interests that includes participatory elements. The state has a responsibility to resolve the above-mentioned dependencies of the transport system on the economic system.

Political decision-makers are faced with the complex task of legitimising transport-related measures as a decision between shaping the future and democratic self-determination in connection with necessary mobility.

An understanding of cause and effect is required if sustainable mobility is to be achieved.

It is essential to differentiate between mobility and transport. Only then is it possible to define information, procedures, measures and objectives.

Transport systems as passenger transport or freight transport are to be divided into their functions in structure, transport process and people (cf. SCHWEDES-RAMMERT 2021,313-314).

- Structure as a spatial element and connection to Traffic movements (think of railway lines, airports and waterways) > Infrastructure
- Traffic process as a temporal change of location in a traffic movement > Traffic and
- People as actors in time, space and type of movement in mobility behaviour, choice of route and choice of means of transport > Mobility.

13.2 Modern transport policy

A transport policy provides the normative framework. This includes spatial, environmental, technological, economic and social requirements (cf. SCHWEDES 2018, 3-24).

The guiding principle of economic development has dominated for a long time. In the meantime, social and ecological sustainability in conjunction with a contemporary transport policy is gaining in importance.

Mobility opens up new opportunities to influence transport and its effects before they arise. The paradigm shift is changing the political organisation of the transport sector.

Exemplary instruments such as taxes, subsidies, laws and increased information and education can be used to ensure added value for society as a whole.

Networks with projects, regional educational measures and, consequently, national and international institutions for implementation should not be overlooked.

In this context, state intervention for needs structures can/should be particularly useful in the transport sector (cf. DÖHLER 2006, 208-227). In the case of private transport systems, economic interests take centre stage and mobility-related added value is at best a marginal product.

This also corresponds to the provision of services of general interest in supranational treaties, which is enshrined in the Federal Constitution and international law (cf. SCHWEDES-RAMMERT2021.316).

Learning cultures in continuing education

Educational tasks in civic education for achieving the objectives - in the context of "lifelong learning" - as necessary factors for sustainability require an examination of the learning cultures in further and continuing education measures (cf. DICHATSCHEK2018, 20-29).

It appears that there is a slow transition from "adult education" to "adult learning". The interest relates to age-related learning styles, knowledge transfer, adult-typical learning experiences, age-related learning styles and learning requirements (cf. NOLDA2008, 81-94).

The areas of change, experience and reflection are essential to the concept of learning.

- It can be assumed that there is no evidence of a general decline in cognitive abilities before the age of 60 (cf. FALTERMAIER-MAYRING-SAUP-STREMEL 2002, 22).
- The intellectual capacity between age groups should not be measured against the performance profile of younger adults. Older people generally need more time to solve set tasks and are usually more precise in their execution than younger people (cf. NOLDA2008, 83).
- The level of education and the requirements of the world of work and the professional world must also be taken into

account when making comparisons.

- The distinction between crystallised intelligence and fluid intelligence is significant (cf. KRUSE-RUDINGER 1997, 50).
    - Crystallised intelligence means the shaping of thinking and knowledge (knowledge content for a society and culture in which one lives), measured for example in language, communication skills and professional knowledge.
    - Fluid intelligence encompasses biological learning capacity, measured for example in non-cultural tasks such as pictorial representations and simple symbols.

14 Learning ability - Learning techniques

14.1 Learning research

Edward THORNDIKE's "Adult Learning" (1928) is a classic of empirical learning research.

In 1924, comparisons of the test results of younger and older adults in typing and Esperanto, the learning success of older people in evening classes and the learning performance of 15 to 50-year-old adults showed that older people need less time to learn the subject matter, but are slower to learn than normal learners. Younger people were superior in motor skills.

The overall result for the older employees was positive. "Almost anything can be learnt up to the age of 50. Biological age is less important than practice and motivation" (SIEBERT 2012, 18).

Skills that are not or rarely used are forgotten. Interest in learning declines with age. A distinction must be made between different learning achievements and influencing factors.

In his 1961 book "The Inquiring Mind", Cyril HOULE distinguishes between three types of learners: "goal-oriented learners", "activity-oriented learners" and "learning-oriented learners" (cf. KNOLL 2007, 394). They refer to the pursuit of professional goals, interest in social contacts and personal learning as learning goals.

Richard OLECHOWSKI (1972) refers to "learning ability" as an abstract theoretical construct that cannot be precisely determined.

At best, learning achievements can be measured, but these are also complex. Factors include motivation, mental health, reaction latency, accuracy and learning speed, regular training of mental performance, transfer effects from previous learning processes and interference (cf. OLECHOWSKI 1972, 207). We must warn against generalised theses.

14.2 Further training

In continuing education and adult education, the main focus is on didactic considerations.

This includes the principle of participant-orientation and the concept of the teacher as a "facilliator" (supporter of learning/"learning coach").

These principles are used in the area of personnel development (cf. KNOWLES - HOLTON - SWANSON 2007, 149 - 165). Knud ILLERIS' "learning triangle" model is reflected in > cognition - learning content > emotions > motivation (cf. ILLRIS 2006, 29-41; NOLDA 2008, 85-86).

In Klaus HOLZKAMP's subject-scientific learning theory, the learner is addressed in his or her individual learning rationale in a process of appropriation and orientation. A problem of action becomes a problem of learning with the help of a learning technique that is close to life and includes a "learning loop" of experiences ("expansive learning")

- The object of learning has social significance and is subjectively realised by the individual.
- In the case of learning resistance from school experience, opportunities in political education are utilised in virtual learning and in company projects (cf. FAULSTICH - ZEUNER 2006, 31). The low numbers of participants in adult political education are explained by learning conditions that are similar to those at school (cf. NOLDA2008, 88).

14.3 Socialisation

Learning cannot be overlooked as a socialisation process. Adult learning has received new impetus from socialisation research.

Examples include the work of Orville BRIM, Stanton WHEELER

(1974), Hartmut GRIESE (1979), Knud ILLERIS (2004), Heinz BARZ - Rudolf TIPPELT (2004) and Horst SIEBERT (2012).

- BRIM - WHEELER (1974) Change in knowledge and behaviour, but hardly in skills, values and motivation; less new knowledge that is "synthesised" (cf. BRIM-WHEELER 1974, 30). However, new knowledge is important to increase connectivity and sustainability.
- GRIESE (1979) links adult socialisation with "lifelong learning"; the continuity of the life course must be taken into account, sometimes combined with breaks, caesuras and switching points, often with learning requirements (changes - "transformations") in attitude, motivation and value orientation (cf. GRIESE 1979, 187).
- ILLERIS (2004) attempts four phases of a learning biography (cf. ILLERIS 2004, 208, 215, 219, 222 > childhood, adolescence, adult with goals and "harmony")
- SIEBERT (2012) emphasises social changes, the individualisation of life courses and the dissolution of normal biographies, which make no sense in terms of phasing (cf. SIEBERT 2012, 22).

- BARZ- TIPPELT (2004) deal with milieu research, which since the 1990s has concerned learning research alongside class-specific social research with social structures and social status based on school education, income, professional position, mentalities and the pluralisation of lifestyles.
    - It is about milieu-specific school qualifications, learning biographies, lifestyles, learning interests and learning motives with educational topics (cf. also SIEBERT 2012, 26).
    - However, this needs to be relativised. Cross-milieu characteristics such as gender, illness and age sometimes do not result in a milieu categorisation

15 Forms of learning

Independent, hidden learning and learning "just in time" are discussed below. The meaning refers to learning that is only partially influenced by educational institutions and learning and education at this age also takes place outside of institutions.

15.1 Independent learning

Independent learning as a form of learning consists in an individual self-confidence, sometimes in an association with bad school experiences, self-control euphoria, an increase in autonomy and self-organisation.

However, despite subjectivisation, there is a dependency similar to classical learning culture in a conceptualisation (cf. FLEIGE 2011, 51). Just think of the various forms of distance learning in their conceptualisation.

15.2 Hidden learning

European education policy documents distinguish between formal, non-formal and informal learning as core forms.

Depending on the training and educational institution - institution, workplace, leisure time - everyday life, family, structure according to learning times, they designate forms of learning (cf. EUROPEAN COMMISSION 2001).

This addresses hidden learning and acquired skills, for example for voluntary work ("volunteering") and also recognition in application procedures.

The issue of certification procedures and recognition in the Austrian National Qualification Framework (2016) remains open.

Sustainability means national and Europe-wide recognition of knowledge, skills and competences.

15.3 Learning "just in time"

Just-in-time" learning addresses new media and networks as forms of learning.

It requires a learning attitude to be ready to learn.

This leads less to "stockpiling learning" and more to selective learning. Fleeting and diverse knowledge becomes a matter of course in an ongoing form of learning (cf. NOLDA2008, 93).

16 Disruptive factors

They cause serious problems in drop-out phenomena in several respects.

Economic factors such as the cost of the offer, the image of the facility and the benefits of the offer, organisational aspects such as the organisation of the offer, pedagogical aspects such as the atmosphere, group structure, excessive or insufficient demands and individual aspects such as experiences of failure, motivation, perseverance and learning requirements can all play a role.

A distinction must be made between participant fluctuation ("dovecote atmosphere"), a course cancellation ("drop out") and a loss of participants due to ever decreasing demand (cf. SIEBERT 2012, 34).

Coincidences or small things in the form of annoyance or disappointment, but also career changes or sudden illness can lead to cancellation. The greater the interest in an event and the greater the social needs, the more likely it is to be attended to the end.

The time commitment should be correctly assessed. Participants with a higher level of education are more likely to drop out because they believe they can also work on the content at home.

As a rule, participation is based on attractive factors such as encouragement, enjoyment and interests as well as aversive factors such as stress. There should be a balanced weighting between attractive and aversive factors, as changes usually occur during the course of the event.

Pedagogical measures are therefore the clarity of the event announcement, the appropriateness of the event times with an explanation of the time required, a harmonisation of the educational offer and the participant requirements.

Discussions on expectations should be held at the beginning.

In the event of absence, participants should be given the opportunity to connect in terms of content. Competitive thinking and rivalries must be prevented.

Business ethics - Environmental ethics

Ethics is understood

- as motivation to act, which is related to the respective contexts.
- This means a reference to the world, deepened by sociological reflections.
- Depending on the respective field of action, a reference science is required for professional and ethical judgement.

In this context, the term "applied ethics" is often used. The term is intended to describe ethics as a social technology to be applied in good behaviour. In reality, areas such as the environment and the economy are places that actualise the question. It is primarily about normative reflections in questions of decisions and judgements.

When ethics reflects on areas of human life, one should speak of "domain ethics". Social changes cannot set specific standards, otherwise political, socio-economic and cultural change would be rejected.

Due to the increasing economisation of the world we live in, the area of business/economics is of particular interest for school and adult education.

Business ethics

17 Integrative business ethics

Integrative business ethics seeks to uncover and reflect on the normative in economic factual logic. The economy based on the division of labour serves society to satisfy the needs of the standard of living and quality of life (cf. ULRICH 1998, 11).

- It serves the purpose of life.
- The logic of material constraints creates a contradiction to the guiding principle of a good life or just coexistence (cf. protection of the environment alongside increased pressure to perform or high economic growth with consumer prosperity for some people).

- At this point, where economics seems to manage without an ethical dimension, Peter ULRICH calls for a critical clarification of the relationship between economics and ethics on the basis of autonomy (cf. ULRICH 1998, 24).
- ULRICH emphasises the necessity of a post-conventional standpoint of morality that transcends the ethical experience of the great world cultures and world religions towards a culturally invariant standpoint. This morality can only be based on man's self-critical reason as an ethics of reason (cf. ULRICH 1998, 42).
- For such a universal ethic, ULRICH names interpersonal obligations (cf. ULRICH 1998, 44). These are regarded as indisputable basic principles (equal vulnerability of all human beings, the ability to empathise with others, reciprocity of moral claims and the principle of universalisation; cf. WEISS 1997, 217-243).

From a rational-ethical point of view, the factual logic of the (alleged) regulatory function of the market is to be analysed for its validity and the dichotomy of ethical reason and economic rationality integrated into a common perspective of a life-serving economy. The aim is an economic ethics without the respective conditions of the market economy. Accordingly, economic reason is subjected to a critical view.

18 Criticism of economism

Economism is described as the belief of economic rationality in nothing but itself (cf. ULRICH 1998, 127).

- It is based on the thesis of an independent economy, the absolutisation of cost-benefit thinking and a normative exaggeration of the market (cf. the economic logic of the mutual exchange of benefits).
- The idea of efficiency (maximising utility) becomes a basic norm (cf. total market economy).
- The necessity of an economic imperative with a service to the common good is asserted.

ULRICH blames the economic liberalism of the 19th century, which asserted the meaningfulness of the market and is today enhanced by rational economism, for the constraint.

The constraint situation can be changed either through changes in personal preferences (personal self-limitation) or in terms of institutional ethics by limiting competition with individual ethical consequences. Autonomous ethical decisions and intentions on the part of market participants are required.

Environmental ethics

19 Environment

The term means a humane organisation of the living space in connection with a developed identity of a way of life (cf. the concept of home). The need for environmental protection arises from the impact of human intervention.

Environmental protection is more than just nature conservation, but rather the protection of people from the inherent dynamics and lack of overview of the structural consequences of the pursuit of economic profit in a market economy (see SCHWEIDLER 2018, 160161).

20 Principles of environmental protection

The following three principles result from the relationship between a fair distribution of benefits and burdens between people

- Polluter pays principle - damage must primarily be repaired or compensated by the polluter, and secondarily by the state community as the bearer of responsibility for the polluter's legal status.
- Precautionary principle - damage cannot be measured in terms of costs in the same way as economic material wear and tear; rather, it must be avoided as a precautionary measure, taking into account the well-being of future generations.
- Principle of cooperation - damage occurs across borders and therefore cannot always be attributed to specific economies. An objectively required ethical duty arises from international cooperation to combat them.

Bibliography III

The titles used and/or directly quoted in the article are listed below

Agenda 21. United Nations Conference on Environment and Development Rio de Janeiro, June 1992 > http://www.un.org/Depts/german/conf/agenda21/agenda 21.pdf. (11.10.20)

Authors' group on specialised didactics (2011): Konzepte der politischen Bildung, Federal Agency for Civic Education, publication series vol. 1141, Bonn

Barz H. - Tippelt R. (2004): Continuing Education and Social Milieus in Germany, Vol.1-2, Bielefeld

Baumgartner Chr. (2021): Sustainable tourism development. Experiences from rural - alpine regions, in. Brandl St. - Berg W. - Herntrei M. - Steckenbauer Chr./ Lachmann - Falkner S. (eds.): Tourism and rural areas. Innovative strategies and instruments for shaping the future, Berlin, 13-32

Beck U. (1986): Risk Society. Auf dem Weg in eine andere Moderne, Frankfurt/ M.

Beck U. (1998): World risk society. In search of lost security, Frankfurt/M.

Behringer W. (2011): Cultural history of the climate. From the ice age to global warming, Munich

Bockhorst H. - Reinwand V.J. - Zacharias W. (2012): Handbook of Cultural Education, Munich

Brim O.-WheelerS. (1974): Adult socialisation, Stuttgart

Federal Ministry of Education (2016) > https://www.unesco.at/fileadmin/Redaktion/Publikationen/Publikations- Dokumente/2016 Unterrichtsmappe fuer die 5.bis 9.Schulstufe.pdf (22.3.2020)

Federal Ministry of Health and Women's Affairs (2012): Health Goals Austria. Trend-setting proposals for a healthier Austria, 2017 edition, Vienna

Federal Agency for Civic Education (2006): Economy today,

publication series vol. 499, Bonn

CCCA Climate Change Centre Austria - Climate Research Centre Austria Vienna (ed.) (2018): Austrian tourism in a changing climate, Graz

de Haan G. (1999): On the foundations of "Education for Sustainable Development" in schools, in: Unterrichtswissenschaft3/1999, 252-280

de Haan G. (2004): Political education for sustainability, in: Aus Politik und Zeitgeschichte 7-8/2004, 39-46

de Haan G. (ed.) (2007): Study and Research on Sustainability, Bielefeld

Holzmann S. - Petersen D. - Wortmann M. (2022): Sustainable market economy. Economic policy goals and conflicting objectives in the ecological transformation, Gütersloh

Dichatschek G. (2017a): Didactics of civic education. Theory, practice and fields of action in the didactics of civic education, Saarbrücken

Dichatschek G. (2017b): Interculturality. A contribution to theory, education and fields of action in the context of intercultural openness and civic education, Saarbrücken

Dichatschek G. (2018): Learning cultures in adult and continuing education. A contribution to theory, practice and action-specific challenges in the context of civic education, Saarbrücken

Dichatschek G. (2019): Teaching at the university. A contribution to dimensions of teaching, teacher education, further education and training of teachers and university didactics, Saarbrücken

Dichatschek G. (2020): Teacher training - Theory and practice of professionalising the training of teachers, school leaders and school quality management, Saarbrücken

Dollinger Fr (2021): The dilemma and paradoxes of spatial planning, Vienna

Döhler M. (2008): Regulative Politik und die Transformation der klassischen Verwaltung, in: Politische Vierteljahreszeitschrift, Sonderheft 37/2006, 208227

European Commission (2001): Communication from the Commission. Making a European area of lifelong learning a reality. Brussels/ 21.11.2002, COM (2001,678

Protestant Distance Learning Centre (2013): Distance learning course Sustainable Development - Introductory booklet, Study letter 1 Justice, Study letter 2 Peace, Study letter 3 Creation, Study letter 4 Future, Frankfurt/ M.

Protestant Church in Austria (2022): Policy paper of the General Synod "Faith in Creation in the Climate Crisis" > epdö/ Synode/Klimaschutz/ Kirchen v. 10.12.2022 > https://archive.newsletter2go.com/?n2g=cc35u22b-g39z6tuq-k13

Faltermaier T. - Mayring P.- Saup P. - Stremei P. (2002): Developmental psychology of adulthood, Stuttgart

Famulla G. - E.- Butz B. - Deeken S. - Michaelis U. - Möhle V.- Schäfer B. (2008): Vocational orientation as a process. Promoting personality, developing school, ensuring transition, vol. 5 School - Economy/ Working life, Hohengehren

Faulstich P. - Zeuner Chr. (2006): Adult education. An action-orientated introduction to theory, didactics and addressees, Weinheim

Fleige M.(2011): Learning cultures in public adult education. Theory-developing and empirical observations using the example of Protestant organisations. Internationale Hochschulschriften, vol. 554, Münster - New York - Munich - Berlin

Freericks R. - Hartmann R. - Stecker B. (2010): Leisure Science. Handbook for pedagogy,. Management and Sustainable Development, Munich

Frey K. (1991): The project method, Weinheim - Basel

Gallus A. - Jesse E. (eds.) (2007): Forms of government. Von der Antike bis zur Gegenwart, Federal Agency for Civic Education, Schriftenreihe Bd. 624, Bonn

Gardener Cl. (2020): Climate, Corona and Christianity. Religious education for sustainable development in a wounded world, Bielefeld

Griese H. (1979): Adult socialisation research, in: Siebert H. (ed.):

Taschenbuch derWeiterbildungsforschung, Baltmannsweiler, 172-210

Grober U. (2010): Die Entdeckung der Nachhaltigkeit: Kulturgeschichte eines Begriffs, Munich

Grunwald A. - KopfmüllerJ. (2012): Sustainability, Frankfurt/ M. - New York

Gütl Br.- Orthey F.M.- Laske St. (eds.) (2008): Bildungsmanagement. Forming differences between system and environment, Munich and Mering

HauffV. (1987): UnseregemeinsameZukunft, Greven

Hirschmann K. (2017): The rise of national populism. How Western societies are being polarised, Federal Agency for Civic Education, publication series vol. 10150, Bonn

Illeris K. (2004): The three Dimensions of Learning, Roskilde

Illeris K. (2006): The "learning triangle", framework concept for an overarching understanding of human learning, in: Nuissl E. (ed.): Vom Lernen zum Lehren. Learning and teaching research for continuing education, Bielefeld, 29-41

Keeley Br. (2010): International Migration. The human side of globalisation, Bundeszentrale für politische Bildung, Schriftenreihe Bd. 1060, Bonn

Klein E. (2020): The Deep Rift. The History of the Divided States of America, Hamburg

Knoll J. (2007): The Inquiring Mind, in: Koerrenz R.- Meihammer E.- Schneider K. (eds.): Wegweisende Werke zur Erwachsenenbildung, Jena, 384-400

Knowles M.S.- Holton E.- Swanson R.A. (2007): Lifelong learning. Andragogy and adult learning, Munich

Kost A.- Massing P. - Reiser (eds.) (2020): Handbook of Democracy, Frankfurt/M.

Kraatz Cl. (2021): Staying sustainably efficient, Wiesbaden

Kruse A. - Rudinger G. (1997): Lernen und Leistung im Erwachsenenalter, in: Weinert F.- Mandel H. (Eds.): Psychologie der

Erwachsenenbildung, Göttingen, 45-85

Levitsky St. - Ziblatt D. (2018): How democracies die. And what we can do about it, Federal Agency for Civic Education, publication series vol. 10326, Bonn

Luger K. (2012): Sustainability considerations for Salzburg Tourism, in: Land Salzburg (ed.): Enquete des Landtages am 9. Oktober 2012. Salzburg: Schriftenreihe des Landes - Medienzentrums 2012, 105-126

Luger K. (2022): Tourism - About travelling and holidays in our time, Wiesbaden

Luhmann N. (1991): Sociology of Risk, Berlin - New York

Matzner M. (ed.) (2012): Handbook Migration and Education, Weinheim - Basel

Meadows D.H. - Zahn E. - Milling P. (1972): Limits to Growth. Report of the Club of Rome on the state of humanity, Stuttgart

Meadows. D.H. - Zahn E.- Milling P. (2008): Limits to Growth. The 30-year update: Signalling a change of course, Stuttgart

Meisch S. (2014): Tübinger Beiträge zur Hochschuldidaktik - Bildung für Nachhaltige Entwicklung. The concept and its potential for traditional comprehensive universities, Vol. 10/1, Centre for University Didactics, University of Tübingen, Tübingen

Nicklas H. - Müller B. - Kordes H. (eds.) (2006): Thinking and acting interculturally. Theoretische Grundlagen und gesellschaftliche Praxis, Federal Agency for Civic Education, Schriftenreihe Bd. 595, Bonn

Nolda S. (2008): Introduction to the theory of adult education. Grundwissen Erziehungswissenschaft, Darmstadt

Ott K.- Döring R. (2008): Theory and practice of strong sustainability, Marburg

Pufe I, (2017): Sustainability, utb 8705, Constance - Munich

Rademacher F.J. - Beyers B. (2011): World with a future: The eco-social perspective, Hamburg

Reckwitz A. (2019): The end of illusions. Politics, economics and culture in late modernity, Berlin

Rein H. - Stradas W. (eds.) (2015): Sustainable Tourism, Constance - Munich

Rogall H. (2003): Actors of sustainable development, Munich

Rogall H. (2008): Ecological Economics. An introduction, Wiesbaden

Scheidig F. (2016): Professionalism of political adult education between theory and practice, Bad Heilbrunn

Schweidler W. (2018): A brief introduction to applied ethics, Wiesbaden

Schuhmacher E.F. (1973): Small is Beautiful: Economics as if People Mattered, New York

Schwedes O. (2018): Transport policy. An interdisciplinary introduction, Wiesbaden

Sweden's 0.(2021): Public mobility. Prerequisites for people-centred transport planning, Wiesbaden

Schwedes O.- Rammert A. (2021): Transport Policy, in: Siebenpfeiffer W. (ed.): Mobility of the Future. Intermodal transport concepts, Berlin, 312-319

Schweigier W. (2018): A brief introduction to applied ethics, Wiesbaden

Segrist D. - Gessner S. - L. Ketterer Bonnelame (2015): Nature-orientated tourism. Quality standards for gentle travelling in the Alps, Bern

Siebenhaar H.P. (2017): Austria. Die zerrissene Republik, Federal Agency for Civic Education, publication series vol. 10113, Bonn

Siebenpfeffer W. (ed.) (2021): Mobility of the future. Intermodal transport concepts, Berlin

Siebert H. (2012): Learning and education of adults, Bielefeld

Stern N. (2007): The Economics of Climate Change. The Stern Review. Cambridge Press Cambridge

Thorndike E. (1928): Adult learning, New York

Ulrich P. (1998/2001): Integrative economic ethics. Foundations of a life-serving economy, Bern

Federal Environment Agency Sustainable mobility > https://www.umweltbundesamt.de/print/3316 (11.6.2022)

UNESCO (2007): The 2009 UNESCO Framework for Cultural Statistics (Draft) Online > http://www.uis.unesco.org/template/pdf/cscl/framework/draftdocEN.pdf (15.10.20)

Association of the Bavarian Economy (ed.) (2021): Sustainability in education - what needs to be done now. Expert opinion, Münster

Virtual Academy for Sustainability - University of Bremen > https://www.va- bne.de/index.php/en (22 March 2020)

Volkmer M. - Werner K. (Eds.) (2020): The Corona Society. Analyses of the situation and prospects for the future, Bielefeld

White U. (1997): Human Dignity/Human Rights. Normative basic order for a global policy, in: Lüttersfelds W. -Mohrs Th.(Eds.): Eine Welt - eine Moral? A controversial debate, Darmstadt, 217-243

Zoche P. - Kaufmann St.- Haverkamp R. (eds.) (2011): Civil Security. The social dimension of current security policies, Bielefeld

To the author

APS teaching degree (VS - HS - PL 1970, 1975, 1976), certified student counsellor (1975) and school development counsellor (1999), member of the teacher examination commission for the APS at the Tyrolean Regional School Board (1993-2002).

Graduate of Höhere Bundeslehranstalt für alpenländische Landwirtschaft Ursprung - Klessheim/ Reifeprüfung, Maturantenlehrgang der Lehrerbildungsanstalt Innsbruck/ Reifeprüfung - Studies in Educational Science/ University of Innsbruck/ Doctorate (1985), 1st Ecumenism course - Kardinal König Akademie/ Vienna/ Certification (2006); 10th University course in Political Education/ University of Salzburg - Klagenfurt/ MSc (2008), Continuing Education Academy Austria/ Vienna/ Diploma (2010), 6th University course in Intercultural Competence/ University of Salzburg/ Diploma (2012), 4th Internal course in Higher Education Didactics/ University of Salzburg/ Certification (2016). University course Intercultural Competence/ University of Salzburg/ Diploma (2012), 4th internal course University Didactics/ University of Salzburg/ Certification (2016) - Distance learning course Basic Course Adult Education/ Protestant Distance Learning Centre, Comenius Institute Münster/ Certification (2018), Distance learning course Sustainable Development/ Protestant Distance Learning Centre, Comenius Institute Münster/ Certification (2020).

Lecturer at the Institute for Educational Science/ University of Vienna/ Vocational Education - Pre-Vocational Education VO - SE (19902011), Department of History/ University of Salzburg/ Teacher Training Programme History - Social Studies - Political Education - SE Didactics of Political Education (20262017).

Member of the Education Commission of the Evangelical Church of Austria (20002011), sfv. *Director of the* Protestant Education Centre Tyrol (2004 - 2009, 2017 -2019).

Course leader at the VHSn Salzburg Zell/ See, Saalfelden and City of Salzburg/ "Freude an Bildung" - Political Education (2012-2019).

MAIL dichatschek (AT) kitz.net

MIX
Papier aus verantwortungsvollen Quellen
Paper from responsible sources
FSC® C105338

Printed by Books on Demand GmbH, Norderstedt / Germany